Table of Contents

I0468341

Date of Publication: February 2016
ISBN-13: 978-1523818822
ISBN-10: 1523818824
Library of Congress Control Number: 9781523818822
CreateSpace Independent Publishing Platform
North Charleston, SC

GNU Octave Matlab Tutorial Series Vol.1

Octave/Matlab Primer and Applications
EZ Guide to the Commands and Graphics
Edition 1.0

S. Nakamura

Publisher
CreateSpace
7290 Investment Dr. Suite B
North Charleston, SC 29418, USA

About the image on the front cover

The orbit of Hubble outside Earth's atmosphere allows it to take extremely high-resolution images with negligible background light. Hubble has recorded some of the most detailed visible-light images ever, allowing a deep view into space and time. Many Hubble observations have led to breakthroughs in astrophysics, such as accurately determining the rate of expansion of the universe.

The image on the front cover is V838 Monocerotis (V838 Mon) that is a red variable star in the constellation Monoceros about 20,000 light years from the Sun. The previously unknown star was observed in early 2002 experiencing a major outburst, and was possibly one of the largest known stars for a short period following the outburst. Originally believed to be a typical nova eruption, it was then realized to be something completely different. The remnant is evolving rapidly. By 2009 its temperature had increased to 3,270 K and its luminosity was 15,000 times solar, but its radius had decreased to 380 times that of the Sun although the ejecta continues to expand. The opaque ejected dust cloud has completely engulfed the B-type companion. (Excerpts from Wikipedia)

Preface

This book serves as a hands-on tutorial for beginners who are unfamiliar with Octave/Matlab, and is an expansion of the previous book titled, *GNU Octave Primer for Beginners*. This book inherits the first two chapters of the latter but the contents are expanded so it is applicable to the Matlab users also. Additionally two more chapters are added, one chapter on numerical methods and another on application of Octave/Matlab to numerous utilities and games.

Octave (also called GNU Octave) and Matlab are high-level programming language useful to scientific and business students and professionals as well. Both are great tools in studying and practicing mathematics and data processing. Octave and Matlab are compatible to each other although they are not exactly the same.

Yet one significant difference is that Octave is available free of any charge. If you choose to use Octave, the first thing to do is to download Octave from the following site:
https://www.gnu.org/software/octave/download.html

Octave/Matlab can also be used as a desktop calculator on your PC or Mac or Linux computer. But unlike any ordinary calculator, Octave/Matlab are programmable calculators and have graphic tools which are very useful in studying algebra and calculus.

Octave/Matlab need a little tutorial to begin with. That is the purpose of this writing.

Although scripts (programs) are exchangeable between Octave and Matlab, some minor adjustments may be necessary. This book explains how to make programs perfectly exchangeable between Octave and Matlab.

Octave/Matlab are said to be suitable for numerical methods. That is true, but most of mathematical computations in science and engineering can be fulfilled with Octave/Matlab whether the numerical methods are used or not. Octave/Matlab are also very useful in analyzing and processing financial data. Many users of Fortran, C and C++ find the graphic capabilities of Octave/Matlab useful in post-processing the results of their computations.

The only major drawback of Octave/Matlab is that its processing speed is significantly slower than Fortran, C, or other similar programming languages. However, this is not a great problem unless very lengthy problems such as solving partial differential equations are solved. For such heavily computational project, Fortran, C or other similar language can be used, but the results are nicely post-processed by Octave/Matlab.

For general background of Octave, read the introductory article on Wikipedia: https://en.wikipedia.org/wiki/GNU_Octave

S. Nakamura
Author

Chapter 1
Octave/Matlab Commands and Programming

1.1 Starting up with Octave

After download is completed, Octave-4.0.0-installer.exe can be found in *Download* directory. To install Octave, click on it, and you will be guided by the installation wizard. When installation is finished, icons of Octave-4.0.0(CLI) and Octave-4.0.0(GUI) shown below will appear on the desktop.

Clicking on the former will open CLI, meaning Command Line Interface, and the latter will open Graphic User Interface. We use the former first. Once you learn with CLI, switching to the other is easy as explained later. If you prefer to start with GUI, that is fine too, because GUI has also a command window that is equivalent to CLI. One advantage with the command window of GUI is that copy & paste operation is allowed there, while copy & past is not possible in CLI. The CLI window is illustrated in Figure 1.1.

The >> sign is called *Command Prompt*, after which a small white box is flashing. You can write a command there. Type **date** as the first trial and hit *Return*. Then, today's date is printed like:

ans = 3-Feb-2016

fix(clock)
Likewise if you type **fix(clock)**, the response is

ans = 2016   2   3   10   50   52

which are year, month, day, hour, minute, second, respectively. Also try simply **clock** without writing **fix** to see what happens.

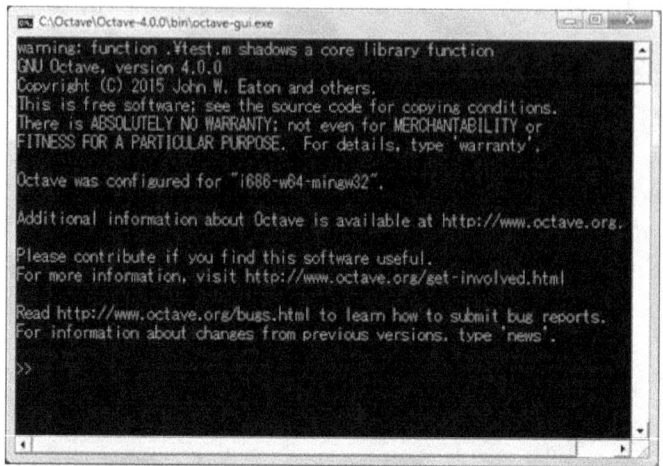

Figure 1.1 CLI

pwd

The most important thing at this stage is to know in what directory (or folder) you are in. To do this, type **pwd** after the prompt. A typical response is

     ans = c:/users/owner

which means currently the directory is **owner** in **users** in **c** drive. Here, **owner** is used as an example, so the directory name **owner** may be different for each PC.

cd

To change the current directory to one of other directories (or folders) contained in the current directory, for example, **sf1**, type **cd sf1** after the prompt >>. Here you can type **pwd** to make sure the directory has been changed. If the answer is

     ans = c:/users/owner/sf1

you are sure that the current directory is **sf1** in **owner** in **users** in **c** drive. To go back to the parent directory **owner**, type **cd** .. (one blank space followed by dot dot). If you want to go to directory **users** while you are in **sf1,** type **cd** ../...   Likewise, to go to **c** drive from **sf1**, type **cd** ../../...

To change directory to another drive like **h** drive (say a stick memory), type **cd h:**. If you have changed to drive **h** but want to come back to **owner** in **users** in drive **c**, type **cd c:/ users/owner.**

mkdir

mkdir creates a directory. **>>mkdir folder_a** creates a directory named **folder_a** in the current directory.

When you work with Octave, you are likely to create many files in the directory. Working in a high level directory such as **users** or **owner** is not desirable. Therefore, it is recommended to make a working directory for each project or period of time by **mkdir** and work in that. Assuming your working directory is to be named **MyOctavePlace,** use **mkdir** as follows:

>>mkdir MyOctavePlace

Then, change directory by

>>cd MyOctavePlace

Assuming **MyOctavePlace** was created in **owner**, make sure that **MyOctavePlace** is in **owner** by >>**pwd**:

ans = c:/users/owner/ MyOctavePlace

Of course, **MyOctavePlace** can be created in a different (parent) directory if you choose so.

Because **c:/users/owner** is the default (or current) directory when Octave is started, you have to change directory manually by >>**cd MyOctavePlace** each time.

However **MyOctavePlace** can be made *home directory* (or default directory) by using **setenv** command. Detail of this procedure is explained in Appendix 3. It is suggested for the beginners to ignore this setting until the reader becomes more familiar to the Octave operations.

1.2 Starting up with Matlab

Starting up with Matlab is essentially the same as Octave. Yet, the default directory and GUI look different. Therefore, this section is written for Matlab users.

When Matlab is started from the menu at the left bottom menu on the computer, the GUI with a command window as illustrated in Figure 1.2 opens.

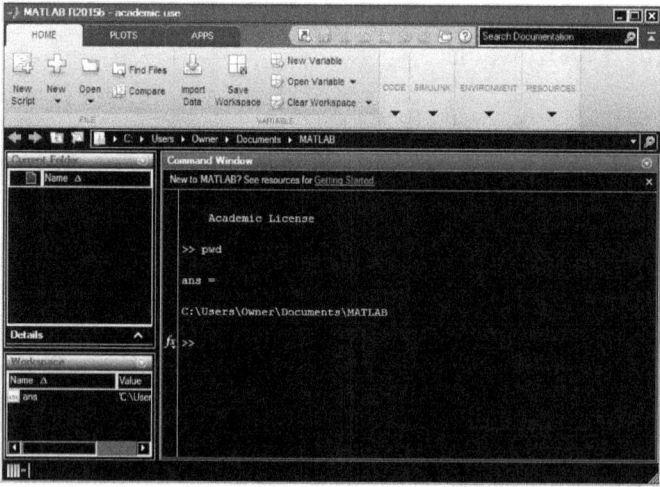

Figure 1.2 Matlab GUI of command window

This GUI has three major sections and two menu bars on the top. The largest section on the right side is the command window.

The **>>** sign is the *Command Prompt*. Type **date** after >> as the first trial and hit *Return*. Then, today's date is printed like:

ans = 03-Feb-2016

fix(clock)
Likewise if you type **fix(clock)**, the response is

ans = 2016   2   3   8   52   34

which are year, month, day, hour, minute, second, respectively. Also try simply **clock** without writing **fix** to see what happens.

pwd
The most important thing at this stage is to know in what directory (or folder) you are in. To do this, type **pwd** after the prompt. A typical response is

ans = C:\matlab_sv12\work

which means currently the directory is **work** in **matlab_sv12** in **c** drive.

mkdir and cd
It is suggested to create a subdirectory within **work** and work normally in that directory. Suppose the name of the subdirectory is **MyMatlabPlace**. To create the **MyMatlabPlace** subdirectory, type **>>mkdir MyMatlabPlace.** To make sure the subdirectory has been created, type **>>dir.** If you see the name of **MyMatlab Place** in the list, the creation succeeded. Now to change the current directory to this new subdirectory, type >>cd **MyMatlab Place.** To make sure you are in **MyMatlabPlace,** type **>>pwd** and see the answer. If the answer is

ans = C:\matlab_sv12\work\ MyMatlabPlace

you are successful. Whenever you open Matlab, use >>**cd My MatlabPlace** and make sure where you are by >>**pwd.**

1.3 Commands you cannot bypass

This section explains the commands that are important to operate the software. Some parts are already explained in the previous sections.

pwd
Shows the current directory name.

dir
Lists the name of subdirectories and files.

cd
Changes the current directory to one of other directories (or folders).

mkdir
mkdir creates a directory. >>**mkdir folder_a** creates a directory named **folder_a** in the current directory.

help
To get some information about a command, use **help**. For example, to get information about the **pwd** command, type **help pwd** after the prompt. The response is as follows:

>>help pwd
       'pwd' is a built-in function from the file
       libinterp/corefcn/dirfns.cc

     -- Built-in Function: pwd ()
     -- Built-in Function: DIR = pwd ()
        Return the current working directory.

See also: cd, dir, ls, mkdir, rmdir.

ls or dir
To know what files and directories (or folders) are in the current directory, type **ls** or **dir** after the prompt.

delete
Command **delete** deletes a file but not a directory. **>>delete file_a** deletes the file named **file_a**.

rmdir
Command **rmdir directoryname** deletes a directory named **directoryname**. However it does not work unless the directory is empty. Therefore, if you wish to delete a directory, you have to delete everything in the directory first.

quit
The **quit** command will close the CLI window, and GUI window as well.

diary on, diary off
With **diary** command, you can record all activities on the command window into a file. The **diary filename** command starts writing all keyboard input and screen output to the file named **filename**. Command **diary off** terminates writing. Once **diary filename** is used, you don't have to write **filename** after **diary**: just write **diary** only. All the recording is appended in the **filename** file, which includes previous diary writings. The file may be opened for reading in the Editor of GUI. If a suffix **txt** is added to the file name, like **filename.txt**, the file may be opened in *Notepad*. Files named like **filename.doc** may be opened in MS Word.

A useful hint: In revising the statement while working on CLI, you do not have to retype the whole things. Use the arrow keys like ↑ and ↓. The lines written earlier will appear, each of which can be revised by first erasing using the *Backspace* key and then rewriting that part and hit *Return*.

13

1.4 Using Octave/Matlab as a powerful calculator

Octave/Matlab can be used as a handy calculator, although it is significantly more powerful than any ordinary calculator. Using Octave/Matlab as a calculator is a good start in learning Octave/Matlab.

Arithmetic operations

Arithmetic operators +, -, *, and / are, respectively, addition, subtraction, multiplication, and division operators, as in traditional programming languages. In addition, ^ is the power operator.

Octave/Matlab has one untraditional operator \ which may be named *inverse division*. This operator yields the reciprocal of division, that is, **a\b** equals **b/a**. For example,

>> c = 3\1
c = 0.3333

Do not use this operator in usual computations, but it will become important when you do linear algebra with Octave/Matlab.

Calculations with scalar variables

As a simple example, let us evaluate the volume of a sphere with radius **r = 2**:

$$\text{Volume} = 4\pi r^3/3 \quad \text{with} \quad r = 2$$

The commands to type are:

>> r = 2;
>> vol = 4*pi*r^3/3;

where **pi** is the name of π, that is, 3.14159265358979 in Octave/Matlab. The caret symbol ^ after **r** is the power operator. The first command **r = 2** is to define the value of **r**. Notice in the preceding script that each line is terminated by a semicolon. The operator **;** is to tell Octave/Matlab to be quiet, that is instructing not to show the results of the command or computation immediately.

When we work in the command window, the computer calculates the answer for each command immediately after the return key is hit. Therefore, the value of **vol** is already in the computer. How can we get the result printed out on the screen?

The quickest way of printing out the result is to type

>> vol

Then, the response is

vol =   33.5103216382911

Of course we could have written as

>> r = 2;
>> vol = 4*pi*r^3/3

where the second line is not terminated by the semicolon. Then, the response is immediately

vol =   33.5103216382911

This number is very long because all the numbers in Octave are in double precision. If you want a short answer, you can tell Octave to print numbers in a short format by

>> format short

After this, the Octave response becomes in the short format like

vol =   33.510

To revert to the long format, the command is

>> format long

If you wish to delete variables like **r** or **vol**, use the command **clear**:

>> clear r
>> clear vol

or

>> clear r vol

Multiple commands (or definitions) can be written in one line, for example,

>> a=1; b=2; c=4;

If each is separated by a comma like

>> a=1, b=2, c=4

then, Octave responds as

a = 1
b = 2
c = 4

Although we did not show examples of using the arithmetic operators such as +, -, *, / and ^ yet, they can be used as needed. For example, $y = x^2 + 3x - 5$ with **x=3** is calculated as

>> x=3; y=x^2 + 3*x − 5
y = 13

The second line above is the answer.

1.5 Calculations involving mathematical functions

Octave/Matlab has a large number of mathematical functions. Some elementary mathematical functions available in Octave/Matlab are shown in Table 1.1.

For more advanced mathematical functions such as Bessel function or Legendre functions, detailed information is available by using the commands

>> help bessel

or

>> help legendre

Table 1.1 Elementary Functions

Trigonometric functions	Remarks
sin(x) cos(x) tan(x) asin(x) acos(x) atan(x) atan2(y, x) sinh(x) cosh(x) tanh(x) asinh(x) acosh(x) atanh(x)	$\pi/2 \geq \text{atan}(x) \geq -\pi/2$ Same as atan(y/x) but $\pi \geq \text{atan2}(x,y) \geq -\pi$
Other elementary functions	Remarks

abs(x)
angle(x)
 	
sqrt(x)
	Absolute value of x
	Phase angle of complex value:
	    if x = real, angle = 0
	    if x = complex, $-\pi/2 < \text{angle} < \pi/2$
	Square root of x

real(x)	Real part of complex value x
imag(x)	Imaginary part of complex value x
conj(x)	Complex conjugate of x
round(x)	Round to the nearest integer
fix(x)	Round a real value x toward zero
floor (x)	Round a real value x toward - ∞
ceil(x)	Round a real value x toward +∞
sign(x)	+1 if x > 0; -1 if x < 0
mod (x, y)	Remainder upon division: x - y*fix(x/y)
rem (x, y)	Remainder upon division: x - y*fix(x/y): different from mod if y ≤ 0
exp(x)	Exponential base e
log(x)	Log base e
log10(x)	Log base 10
factor(x)	Factorizes x into prime numbers
isprime(x)	1 if x is a prime number, 0 if not
factorial(n)	(n)(n-1)(n-2)...(3)(2)(1) if n is a positive integer
sort(x)	Rearrange the numbers in array x to increasing order
sum(x)	Computes the total of the numbers in array x
min(x)	Finds the minim in x
max(x)	Finds the maximum in x
rand('seed', x)	Sets the seed number to x
rand	Generates a random number between 0 and 1
rand(n)	n-by-n matrix of random numbers

The mathematical functions can be used as needed in any calculations, for example:

>> sin(1.2)*exp(1)

which returns the value of sin(1.2)exp(1), namely, ans = 2.5335. The trigonometric functions use radian only. So if a value is in degree, you have to convert it before using the trigonometric functions or in the arguments in the trigonometric functions like

>> sin(angle_1*pi/180)*exp(named_variable)

1.6 Variables and variable names

Variable names and their types do not have to be declared. This is because any variable can take real, complex, and integer values in double precision.

In principle, any name can be used as long as it is compatible in Octave/Matlab. We should, however, be aware of two incompatible situations. The first is that the name is not accepted by Octave/Matlab. The second is that the name is accepted, but it destroys the original meaning of a reserved name. These conflicts can occur with the following types of names:

(a) Names for certain values
(b) Function (subroutine) names
(c) Command names

A bad example of the second conflict is as follows: If **sin** and **cos** are used as user-defined variables, for example,

```
>> sin = 3;
>> cos = sin - 2;
```

the calculations proceed; however, **sin** and **cos** can never be used as trigonometric functions thereafter until variables are cleared by **clear** or Octave is shut down.

Traditionally, symbols i, j, k, 1, m, and n were used as integer variables or indices. At the same time, i and j are used to denote a unit imaginary value, $\sqrt{(-1)}$. In Octave, **i** and **j** are reserved as unit imaginary value. Therefore, if the computation involves complex variables, it is advisable to avoid **i** and **j** as user-defined variables or indices. (Because i and j are so commonly used as indices in mathematics, it is hard not to use these as indices at all. The author uses these as indices in programs that never involve imaginary or complex variables. So you may see such in this book.)

Table 1.2 lists reserved variable names that have special meanings.

In order to examine if a variable or function name you consider is in any conflict, use **exist**. For example, for >>**exist atan**, the response is

   ans = 5

where 5 means that **atan** is a built-in function, so it cannot be used as your user-defined name.

In general, >>**exist name** is responded as one of the following cases:
if **ans**=1, **name** is a variable;
if **ans**=2, **name** is an absolute file name, an ordinary file in Octave/Matlab's 'path', or (after appending '.m') a function file in Octave/Matlab's 'path';
if **ans**=3, **name** is a '.oct' or '.mex' file in Octave/Matlab's 'path';
if **ans**=5, **name** is a built-in function;
if **ans**=7, **name** is a directory;
if **ans**=103, **name** is a function not associated with a file (entered on the command line);
if **ans**=0, **name** does not exist.

Therefore, if **ans**=0, the **name** can be safely used as a variable or file name or function name. Special numbers and variable names are shown in Table 1.2.

Table 1.2   Special Numbers and Variable Names

epsilon: machine epsilon= 2.22204e-16
pi: π = 3.141592653589793
i and j: unit imaginary = $\sqrt{(-1)}$
inf: infinity, ∞
nan: not a number
date: date
clock: clock
flops: floating point operation count
nargin: number of function input count
nargout: same for function output

String variables

Any valid variable names can be used to represent a string. For example,

>> s ='My name is Tom Brown'
s = My name is Tom Brown

Here, the variable **s** is a string variable, which is an array of characters. More details of string variables are explained near the end of Section 1.6.

A command or a sequence of commands may be expressed as a string variable. The commands in the string form can be executed by **eval**. For example,

>> s = 'a=5.33;b=2.2;y=sin(a)*exp(-b)'
>> eval(s)

or more directly

>>eval('a=5.33;b=2.2;y=sin(a)*exp(-b)')

or

>>eval('a=5.33;'); eval('b=2.2; ');
>>eval('y=sin(a)*exp(-b)')

yields the results of calculating y=sin(a)exp(-b) with the defined values of a and b:

y = -0.090334

This feature may not seem so useful at a glance, but it becomes a powerful means in programming because the string variable may be used as a user-defined input, and also the commands in a string form may be   automatically developed within a program. More about **eval** is explained in Appendices 1 and 2,

Complex variables

A complex variable may be defined by writing i or j next to a number, for example

>>a = 5 + 2i

the response is

a = 5 + 2i

Of course you can write as >>a = 5 + 2*i.

The calculation of

>>f = 1 + 2i; g = 1 - 2i;
>>h = f*g

yields

h = 5

where we assume that the reader is familiar with arithmetics of complex numbers. However, writing as

>>a = b + ci

does not work, where b and c are variable names. You have to write as follows:

>>a = b + c*i

or

>>a = b + i*c

How to find what variables have been used

To find a list of user-defined variable names, use the command

>>who

then a response would be, for example:

Variables in the current scope
a   an   f   g   h   s   t

More detailed information may be obtained by **whos**:

>>whos

which is responded as

Variables in the current scope:

Attr	Name	Size	Bytes	Class
c	a	1x1	16	double
	an	1x12	12	char
c	f	1x1	16	double
c	g	1x1	16	double
	h	1x1	8	double
	s	1x36	36	char
	t	1x1	8	double

Total is 53 elements using 112 bytes

1.7   Looped computation with for/end

Octave/Matlab has for/end and while/end loops. In this section only the former is explained. For quick illustration, the following script calculates $y = x^2 - 5x - 3$ for each of $x = 1, 2, .. 5$ in increasing order of x:

>> for x=1:5

```
        y=x^2 - 5*x - 3
    end
```

Notice that **for** is terminated by **end**. In the first cycle, x is set to 1, and y is calculated. In the second cycle, x is set to 2 (with an increment of 1), and y is calculated. The same is repeated until the calculation of y is completed for the last value of x. In writing the above statements on command line, Octave/Matlab does not issue any new prompt sign >> until **end** is written and *Return* is hit. The response of Octave/Matlab to the foregoing statements is

```
y =   -7
y =   -9
y =   -9
y =   -7
y =   -3
```

If x is to be changed with a different increment, the increment can be specified between the initial and last number as follows:

```
>> for x=1:0.5:5
        y=x^2 - 5*x - 3
    end
```

where the increment is now 0.5. The response is

```
y = -7
y = -8.2500
y = -9
y = -9.2500
y = -9
y = -8.2500
y = -7
y = -5.2500
y = -3
```

Order of calculations in the loop can be reversed as follows:

```
>> for x=5:-1:1
```

$$y = x^2 - 5*x - 3$$
    end

Here, the middle number -1 in 5:-1:1 is the (negative) increment (or decrement) in changing x.

The sequence of x does not have to be constantly incremented or decremented. Any sequence of x values may be used as follows. Suppose the computation is desired to be in the written order of x=2, 0, 15, and 6. The for/end loop may be written as

    >> for x=[2, 0, 15, 6]
        $y = x^2 - 5*x - 3$
       end

where x = [2, 0, 15, 6] looks like an array variable introduced in the next section, but when it appears after **for** in a **for/end** loop, it means x is sequentially set to 2, 0, 15, and 6 in the order written. The result of the calculation is

    y = -9
    y = -3
    y =   147
    y =   3

1.8 Array variables

One-dimensional array variables
One-dimensional array variables are in a column or a row form, and are closely related to vectors and matrices. In linear algebraic computations in Octave/Matlab, a row array is used as a *row vector,* and a column array is a *column vector.* However, arithmetic operations of array and vectors are not the same as mentioned later. The variable x can be defined as a row array by specifying its elements, for example, by

```
>> x = [0, 0.1, 0.2, 0.3, 0.4, 0.5]
```

which is responded as

```
x =
     0.00000   0.10000   0.20000   0.30000   0.40000   0.50000
```

To print a particular element, type x with its index or position. For example, typing >>x(3) yields

```
ans =   0.20000
```

An equivalent way of defining the same x is

```
>> for k=1:6
        x(k)=(k-1)*0.1;
   end
```

or

```
>> for k=0:5
        x(k-1)=k*0.1;
   end
```

The size of array does not have to be pre-declared as it is adjusted automatically. The number of elements of x can be increased by defining additional elements, for example,

```
>> x(7) = 0.6;
```

A row array variable with a fixed increment or decrement may be defined as

```
>> x = 2:-0.4:0
```

It yields

```
x = 2.0000   1.6000   1.2000   0.8000   0.4000   0.0000
```

The definition of a column array is similar to a row array except that the elements are separated by semicolons; for example,

>> z = [0; 0.1; 0.2; 0.3; 0.4; 0.5];

An alternative way of defining the same thing is to put a prime after a row array:

>> z = [0, 0.1, 0.2, 0.3, 0.4, 0.5]';

The prime operator is the same as the transpose operator in the linear algebra, so it converts a column vector to a row vector and vice versa. Typing z as a command yields

```
z =
      0
      0.1
      0.2
      0.3
      0.4
      0.5
```

If a single element of an array **c** is specified assuming **c** has not been used, for example,

>> c(8) = 11;

then, c(k)=0 is assumed for k=1 through 7. Typing c yields

```
c =
      0    0    0    0    0    0    0    11
```

A new command >>c =2.2 changes the definition of **c** and responded as

c = 2.2

which is a scalar variable now. If c is again redefined as

>> c(1:2:7) =5

it is responded as

c=
5 0 5 0 5 0 5

Array variables can be combined, for example:

>>x=[1 2 3]; y=[0.5 0.6 0.7]; z=[x y]

yields

z = 1   2   3   0.5   0.6   0.7

Some entries may be eliminated from an array, for example:

>>z = [z(1:2), z(5:6)]

deletes the 3rd to 4th entries from z:

z = 1   2   0.6   0.7

Two-dimensional array variables

Stacking row arrays of the same size multiple times, a two-dimensional array can be created. Likewise, a two-dimensional array may be created by a row array of column arrays. For example, a=[1, 2, 3, 4] and b=[9, 8, 7, 0] are two row arrays of the same length. A new column array may be written as c=[a; b], which becomes a two-dimensional array. To demonstrate creation of the two-dimensional array on the computer, we set

>> a=[1, 2, 3, 4],   b=[9, 8, 7, 0]

and get

```
a =
    1    2    3    4

b =
    9    8    7    0
```

Now >> c=[a; b] yields

```
c =
    1    2    3    4
    9    8    7    0
```

Here c is a two-dimensional array (or 2-by4 array). Of course, c can be created directly by >> c = [1, 2, 3, 4; 9, 8, 7, 0] .

Combining multiple two-dimensional arrays into one two-dimensional array works if all the arrays have the same height. To show an example, we define another two-dimensional array as

>>d=[0.1 0.2; 0.5 0.6]

Then it can be prepended to c that is previously defined:

```
>>c=[d,c]
c =
    0.10000    0.20000    1.00000    2.00000    3.00000    4.00000
    0.50000    0.60000    9.00000    8.00000    7.00000    0.00000
```

Some columns of c can be deleted, for example:

```
>>c=[c(:,1:2), c(:,6)]
c =
    0.10000    0.20000    4.00000
    0.50000    0.60000    0.00000
```

Looped calculations using array variables

With array variables, repeating the same calculations for different input becomes easy and efficient. For example, suppose we wish to compute the values of $y = x^3 + 2x^2 - x$ for $x = 2.2$, 3.1 and 5.0. A script to do the entire calculations with a for/end loop is

```
>> x=[2.2, 3.1, 5.0];
>> for k=1:3
        y(k) = x(k)^3 + 2*x(k)^2 – x(k);
    end
```

Array arithmetic operators

Using the array arithmetic operators, however, the foregoing calculations in the prior subsection can be written much more compactly as

```
>> x=[2.2, 3.1, 5.0];
>> y = x.^3 + 2*x.^2 – x
```

Here notice the ^ is prepended by the period ". " operator. This operator makes it possible to do the computation of ^ for each member of the array variables.

Likewise, the computation involving multiplication and division for array variables may be written in a compact form using the array arithmetic operators. We consider a sample computation with for/end statements first:

```
>>x=[2.2, 3.1, 5.0];y=[1, 3, 5];z=[1, 1, 2];
>>for k=1:3
        u(k) = x(k)*y(k)+ 2*x(k)^2 –x(k)/z(k);
    end
```

With array arithmetic operators, the above script is equivalently written more compactly as

```
>>x=[2.2,3.1,5.0];y=[1,3,5];z=[1,1,2];
>>u = x.*y + 2*x.^2 – x./z;
```

where .*, ./ and .^ are called respectively array multiplication, array division and array power operators.

Another example:

```
>> x=[1 3 0 1]; y=[1 2 2 1];
>> z=exp(x).*y
```

yields

```
z =
      2.7183    40.1711    2.0000    2.7183
```

The above is equivalent to

```
>> for m=1:4
       z(m)=exp(x(m))*y(m);
   end
```

Array length and array size

Array length can be found by the **length** function: for example

```
>> x=[ 1, 3, 5, 0, 1, 1];
>> n=length(x)
```

yields

```
n =  6
```

Therefore, if a calculation is repeated for all members of an array variable x, a script may be written as,

```
>> for k=1:length(x)
       y(k)=x(k) + x(k)^3;
   end
```

Use of **length** is limited to one-dimensional arrays, but **size** becomes useful for two-dimensional arrays. Let us see how **size** works with the two-dimensional array of c:

>> c =[1, 2, 3, 4; 9, 8, 7, 0];   size(c)
ans =
        2   4

which means the vertical height of c is 2 and horizontal length is 4. **size(c)** itself is an array, so >>f=size(c) yields f(1)=2 and f(2)=4.

String variables revisited

String variables, already explained earlier, are arrays of characters. Therefore, multiple strings can be combined.

Here is an example:

>> a='string1', b=' ', c='string2'
>> d = [a, b, c]

yields

a = string1
b =
c = string2
d = string1 string2

where b is a string variable of one blank space. See also

>> a(2:4)
ans = tri

Here, a(2:4) is second though fourth characters of a.

1.9   Loops with while/end

The **while/end** loops are similar to **for/end** loops. The **for/end** loop has a fixed number of loops (repeats), but the number of **while/end** loops is controlled by a condition specified next to **while**. A program of finding the number of negative elements in c is rewritten with **while/end**:

```
c=0;
x=[-8, 0, 2, 5, 7, 2, 0, 0, 4, 6, 6, 9];
k=1;
while k<=length(x)
    if x(k)<0, c=c+1;
    end
    k=k+1;
end
c
```

Sometimes a loop that can continue infinitely is used, which may be terminated when a certain condition if any is met. The following example is an infinite loop that is terminated only by **break** if the condition **x>x1imit** is met:

```
while 1
    --- some calculation of x here ---
    if x > x1imit, break; end
end
```

Here **1** after **while** means that the while loop is to be continued without any condition, so the loop continues on and on until the **break** statement is hit by a chance. Don't run any infinite loop unless some mechanism to stop the loop is built in. If an infinite loop continues, the only way to stop is to shut down the software or computer, provided that you know how. If this occurs while using a time-shared computer, you will have to ask the computer administrator for help shut down the computer, which can be disastrous to many other users on the computer.

1.10 Programming with Octave/Matlab

Being able to develop programs (or scripts) as m-files is one of the most powerful aspects of Octave/Matlab. Any computational work may be done on the command window, but if the steps are very long or the same computational job is repeated over and over, working only on the command window is tedious and not only inefficient but, in the event of errors, correction is often impossible.

With a program developed for such lengthy computations, mistakes in the program can be corrected easily on a program editor and the program (or script) can be run repeatedly any number of times desired. Some effort to learn (1) how to write a program and (2) how to manage the program in the filing system in the computer is necessary, but its benefit overweighs the pain of learning.

An Octave/Matlab program is a sequence of commands written in a script and saved in a directory as an m-file, which is saved as a file with extension `.m`, or `file_name.m`.

Octave editor

When to write a program, using GUI (graphic user interface) illustrated in Figure 1.3 is recommended, although other way of writing without GUI and running the programs from CLI is possible. A GUI can be opened by clicking on the GUI icon that appears in the desktop screen of a PC.

The GUI has two columns. A small box at near top of the left column shows the current directory, which in Figure 1.3 is **c:/users/owner**. (The current directory can be **c:/users/owner/MyOctavePlace** if you have made it the Home directory of your Octave works. At this point, we assume that this change has not been done.)

The directory can be changed as follows: To go to the parent directory, click on the arrow sign ↑ next to the box showing the current directory. The box under the current directory name lists the names of files or sub-directories in the current directory. So by clicking on one of the subdirectories, the clicked subdirectory opens and becomes the current directory.

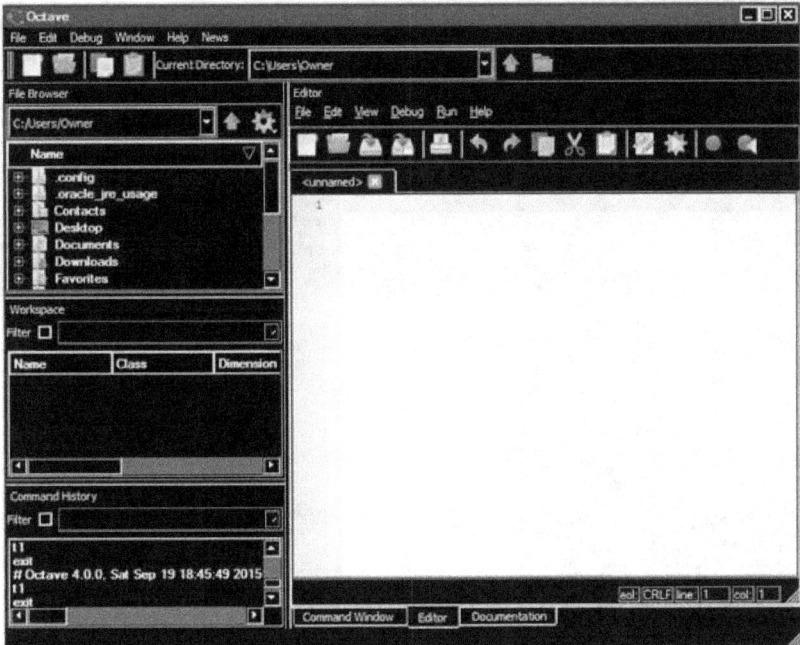

Figure 1.3   Illustration of Octave GUI

Before writing any program, it is recommended to create a new working directory, where you work and save any program you write. The working directory would be in **c:/users /owner**, but if you wish you can choose any appropriate and convenient place. How to create and open a new directory is explained next.

Suppose we create a new directory in the current directory, **c:/users/owner**. By clicking on the gear sign next to the thick green arrow sign, a drawdown menu opens, in which an icon for *New Directory* can be found. Click on that, and a small box opens for writing the name of the new name. After writing the name of the new directory, click OK. Then, the new directory is created and appears in the box of files and subdirectories. Click it to open, and it becomes the current directory. If you have not created the directory **MyOctavePlace**, you may do so now using the procedure just mentioned here.

The right column of GUI contains a large box, which can be *Command Window, Editor*, or *Documentation List*. To see which one is currently on the screen, look at the labels under the large area. In the illustration of the GUI in Figure 1.3, Editor is open. To open another item, just click on the selected label.

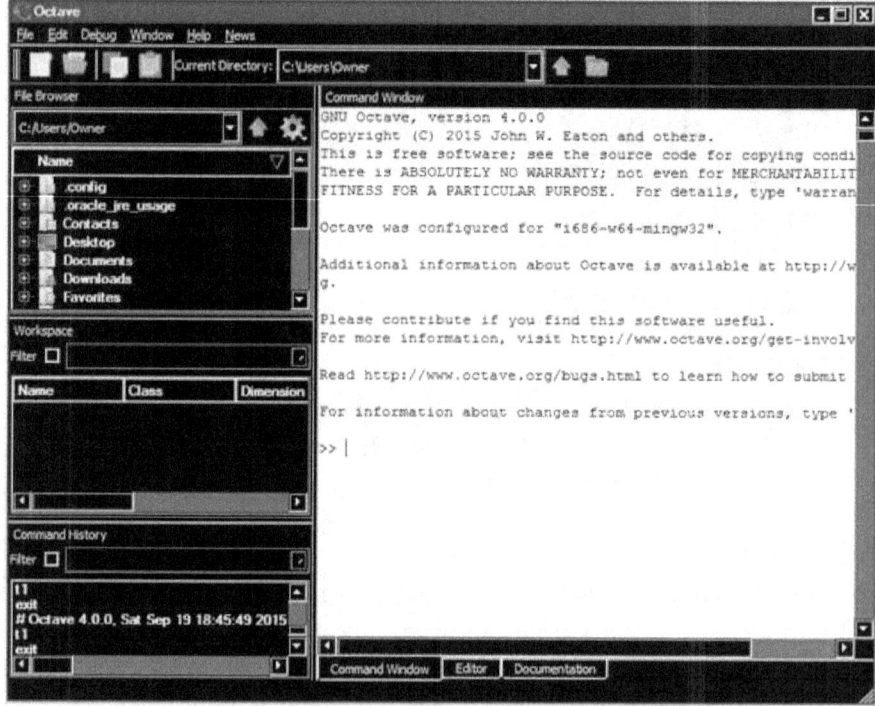

Figure 1.4   Illustration of Command Window of Octave GUI

On the Editor opened, a program can be written with any number of lines. The name of file opened in the Editor is shown at the top left of the Editor. If the name there is unnamed, you will have to give the name before your program is saved and run.

Write a very short program consisting of one or two lines, for example:

```
surf(peaks)
disp('Look at the graphic window to see results.')
```

This script can be run by clicking *Run* in the menu above the editor window. There are two choices in the drawdown menu of Run. The first is *Save File and Run*, and the second is *Run Selection*. If you select the former, you must give the name first, say **prg1**, in the space that opens up. The program created is in the form of **prg1.m** with the suffix **m**. After this, the program starts running if it has no typo or errors. You have to open the Command Window of GUI to read the results because the results can be found only in the Command Window illustrated in Figure 1.4.

If *Run Select* is clicked, a selected part of the program is run without saving the program, where selection means you highlight the part by mouse. This is often a convenient way of testing a part of the program.

The two-line script suggested in the prior subsection will run for sure, but *Run* for your own programs will not succeed in early attempts if there are errors. For the errors, relatively detailed messages appear on the Command Window. Sometimes, however, only a weird message may appear. In the worst cases the whole system of Octave may stop and freeze. If this happens, don't hesitate to close Octave and start again. If Octave does not close, you may have to shut down the PC and restart.

Such a nasty behavior most likely starts when the program has some errors. Therefore, it is advised to write only a short part of the program at a time and run that part by *Run Selection*. When that part runs smoothly, add another small number of lines. Only after the program is thoroughly tested part by part, run the entire program by *Save File and Run*.

Matlab editor

The Matlab editor is slightly different, which opens from the command window (illustrated in Fig. 1.2) by clicking **New** in the draw-down menu at the top menu bar. Figure 1.5 illustrates the editor window of the academic version R2015b, which has four large boxes. On the right side of the window, the large upper box is the editor, while the lower large box is the command window where the results of running of a program is shown. To edit an existing program, click **Open** in the same menu bar and select the file in the directory.

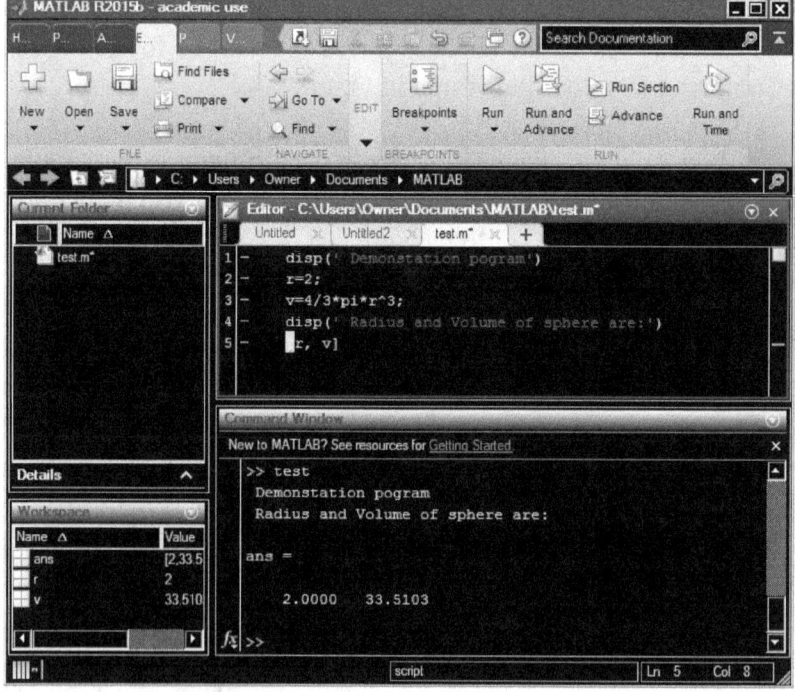

Figure 1.5 Matlab GUI

The appearance of the editor window of Student Version of Matlab may be different, but there is no essential differences from the editor window shown here.

Once the editor window opens, open **File** in the top menu and click **Save As**, and write in the name of this new m-file.

Now a program can be written in the editor as illustrated in Fig. 1.5). To run the program, click open **Run** in the top menu bar, and click **Run:ScriptName**. The results of the run is on the command window, which reads:

> Demonstration program
> Radius and Volume of sphere are:
> ans=
> 2.0   33.5103

The pathological behavior may develop in Matlab also if there is a serious error in the program. If it occurs, quit Matlab and restart.

Echo on, echo off
When an m-file is executed, the statements in the m-file are not usually printed on the screen. After echo is turned on with the **echo on** command, however, the statements are printed. By doing this, the user can see which part of the m-file is being executed. To turn off echo, type **echo off**. This feature may become useful particularly when debugging an m-file.

Comment statements
The percent sign **%** in m-files indicates that any statements after this sign on a line are comments and are ignored in the computations. (One exception is the % sign in format statements.) Also the **disp** command as illustrated in Figure 1.5 is to print out comments on the command window to make reading of output easier. See more about **disp** in Section 1.12.

1.11 Branch operation with if, else, elsif and end

The **if** operator allows switching between two choices or more of actions in the computational flow. An example is shown next:

> if n<= 5, price=15;
>   else price=12;

end

Here we assume that the value of **n** was defined before. Then, if **n** is less than or equal to 5, price is set to 15, and else (otherwise) price is set to 12. The **if** statement is always closed with **end** statement.

The following example is a little more elaborate:

```
if n<= 5, price=15;
    elseif (n>5 & n<10), price=12;
    else price=10;
end
```

Here the second line of **elseif** has been added. The sign & means "and". Therefore, if n is less than or equal to 5, price is set to 15; else if n is greater than 5 and less than 10, price is 12; otherwise (that is, if n is equal to or greater than 10) the price is 10.

Table 1.3 shows all the operators that can be used in if statements.

Table 1.3   Operators associated with if operator

Operators	Meaning
>	Greater than
<	Less than
>=	Greater than or equal to
<=	Less than or equal to
~=	Not equal to
==	Equal to
|	Or
&	And

NOTE: In Octave, && and || may be used instead of & and | respectively. However, Matlab refuses to respect && and || but uses only & and |.

The *not equal* operator is written as "~=":

```
if r ~= 3, vol = (4/3)*pi*r^3;
end
```

Example of *or* operator:

```
if g>3 | g<0, a = 6;
end
```

Example of *and* operator:

```
if a>3 & c<0, b=19;
end
```

Example of *equal* operator: Assuming r and d are integers defined earlier. The following statement finds out if r is divisible by d, or not: if divisible, r is printed out:

```
if fix(r/d) == r/d, r
end
```

or equivalently

```
if mod(r,d) == 0, r
end
```

(See Table 1.1 for the **mod** function.)

The & and | operators can be used in a clustered form, for example,

```
if ((a==2 |  b==3) & c<5) g=1;
end
```

The if/end statement can be inserted in a for/end loop or another if/end loop. In the following example, y=sin(x) takes effect but, if sin(x) becomes negative, y is set to 0:

```
for x=0:0.1:10
    y=sin(x);
    if y<0, y=0; end
    [x,y]
end
```

Here a bit of new style of printing x and y is added, that is [x, y]. In this way both x and y are printed in a single row on the screen.

In the following example, c=0 initializes the counter c to zero, x =... defines an array of numbers, and **length(x)** is the length of the array x. In the **for/end** loop the counter c is incremented by one if x(k) is negative. Finally, c becomes the total count of the negative elements.

```
c=0;
x=[-8, 0, 2, 5, 7, 2, 0, 0, 4, 6, 6, 9];
for k=1:length(x)
    if x(k)<0, c=c+1; end
end
c
```

As another example, we write a program that removes the numbers divisible by 4 from an array x. Assume the array x is initially given by

```
x=[-8, 0, 2, 5, 7, 0, 4, 6, 6, 9, 16];
```

The program for this chore is

```
k=0; clear   y
for n=1:length(x)
    if x(n)/4 ~= fix(x(n)/4),
        k=k+1;
        if k==1, y(1)=x(n);
            else   y=[y,x(n)];
        end
```

```
        end
   end
   x=y
```

Here, y=[y,x(n)] is to append x(n) to y. The result of the run is

$$2 \quad 5 \quad 7 \quad 6 \quad 6 \quad 9$$

==

Remark: In Octave **endif** may be used instead of **end** at the end of if/end loop. Likewise **endwhile** may be used instead of **end** at the end of while/end loop, and **endfor** instead of **end** at the end of for/end loop. The advantage of using **endif, endwhile** and **endfor** is to make the program easier to read, because if only **end** is used, the program may become congested with many **ends**. This practice, however, causes a big conflict if a program developed for Octave is later used in Matlab, because Matlab does not allow **endif, endwhile** and **endfor**.

The author's suggestion to avoid the conflict but still get the same advantage is to use **end%if, end%while**, and **end%for**, because % is an operator to make anything after % a comment so no conflict occurs (except in a format statement).

Another way to reduce confusion is to indent, by 2 blank spaces, each loop of **for/end** , **if/end** , and **while/end**, for example:

```
      for
        while
          if
             for
             end
          end
        end
      end
```

This structured form of writing works well even with **end%for**, **end%if** and **end%while**:

```
for
    while
        if
            for
            end%for
        end%if
    end%while
end%for
```

==

break
The **break** terminates execution of a **for** or **while** loop. When used in nested loops, only the immediate loop where **break** is placed is terminated. In the next example, **break** terminates the inner loop as soon as **n>2*m** is satisfied, but the loop for **m** is continued until the loop of **m=6** is completed:

```
for m=1:6
    for n=1:20
        if n>2*m, break,
        end %if
    end %for
end %for
```

In a programming language that has no break command, **goto** is used to break a loop. Octave/Matlab, on the other hand, have no **goto** command.

1.12   Output technique

Formatted output

In all of the foregoing examples, the results of computations are printed in a very primitive way, that is, just numbers. However, the style of writing can be refined as follows using **fprintf**.

In the following example, the x and y values are printed using **fprintf**:

```
for x=1:4
    y=x^2 - 5*x – 3;
    fprintf('x = %f   y = %f\n', x, y )
end %for
```

where **%f** specifies the format to be the f-format (floating point value format), and **\n** is the line feeder. The result is

```
x = 1.000000   y = 1.000000
x = 2.000000   y = 4.000000
x = 3.000000   y = 9.000000
x = 4.000000   y = 16.000000
```

To see the effect of **\n**, delete it and run to see what happens. You will see the printout is jammed without line feed.

To use the e-format, change **%f** to **%e**:

```
for x=1:4
    y=x^2 - 5*x – 3;
    fprintf('x = %e   y = %e\n', x, y )
end %for
```

The result is

```
x = 1.000000e+000   y = 1.000000e+000
x = 2.000000e+000   y = 4.000000e+000
x = 3.000000e+000   y = 9.000000e+000
x = 4.000000e+000   y = 1.600000e+001
```

When to print an integer, the %i format may be used, for example:

```
z=50
>> fprintf('%i \n', z)
50
```

However, when z is a floating point value, the format is automatically changed to %f:

```
z=pi*100
>> fprintf('%i \n', z)
314.159
```

In using the %f and %e, the number of decimal places may be specified by changing them to, for example, %.3f and %.2e respectively:

```
>> fprintf('%.1f   %.5f \n ', z, pi)
314.2   3.14159
```

```
>> fprintf('%.1e   %.5e \n ', z, pi)
3.1e+002   3.14159e+000
```

disp

Command **disp** displays a number, an array variable, or a string on the command window without variable name. Therefore, it may be used to display messages or data flexibly on the screen. For example, **disp([pi, 2.2, 4.1])** prints

```
3.14159   2.2000   4.100
```

on the command screen.

The **disp** command is useful to warn the user of a program before **input** command is used by instructing how to write the input, for example, disp('The following input must be a string enclosed by single quote signs')

The **disp** command even evaluates an equation, for example

>>x=pi/4; disp(y=sin(x)*cos(x))

calculates the equation, **y=sin(x)*cos(x)**, and sets the value of **y**. If the equation is written directly like >> **y=sin(x)*cos(x)**, "**y=**" is printed before the value of y, but **disp(y=sin(x)*cos(x))** simply displays the y value without "y=". Consider using **disp** as:

x=pi/4;
disp('sin(x)*cos(x)=')
disp(y=sin(x)*cos(x))

The response is

sin(x)*cos(x)=
0.50000

An example of displaying the result of multiple equations is as follows:

x=pi/4; disp([y1=sin(x), y2=cos(x), y3=tan(x)])
0.70711   0.70711   1.00000

sprintf
It is very similar to **fprintf** except that **sprintf** writes the output into a string.

This statement is often used to create a command in a string that can be executed as **feval(s, a)** where **s** is a string of a function name, and **a** is the argument of the function. For example:

>>s=sprintf('tan'); feval(s,1)
ans =   1.5574

It is useful when a command is to be created or edited automatically and executed within an m-file.

The format statements explained for **fprintf** also work with **sprintf**: for example,

```
>> S=sprintf(' %.1f ', pi)
S =   3.1
>> S=sprintf(' %.7f ', pi)
S =   3.1415927
>> S=sprintf(' %.7e ', pi)
S =   3.1415927e+000
```

Writing into a file

With **fprintf**, it is possible to write into a file rather than printing on the screen. To do this, a named file must be opened by **fopen**. The following example illustrates the procedure:

```
vol=55.8;
file_id = fopen('file_x', 'w')
fprintf( file_id, 'volume= %f\n', vol)
fclose(file_id)
```

Here, **file_x** is first opened, and the value of **vol** is written in the file using the f-format. Then the file is closed. The file may be viewed on the GUI Editor if you need to see the contents.

1.13   Input technique

To input a value to the program being run is possible by input. A simple example is

```
x = input('Type input for x and hit Return:   ')
```

The result is

```
Type input for x and hit Return:   3.14
x =   3.1400
```

"Type input for x and hit Return: " is the printout on the command window, and 3.14 is what the user typed on the key board. The next "x = 3.14" is a printout on the command window because the statement of "x = input(...)" is not terminated by the **;** operator.

The input statement can be used anywhere in the script including **for/end%for** and **while/end%while** loops. Here is another example:

```
r=0:
while r<10
    r=input('Type radius (or -1 to stop):');
    if r< 0, break, end %if
    vol = (4/3)*pi*r^3;
    fprintf('Volume = %f\n', vol)
end %while
```

In the foregoing loop, the value of radius r is typed on the keyboard. The **fprintf** statement is to print out vol with the f-format, %f. If 0<r<10, vol is computed and printed out, but if r<0 the loop is terminated. Also, if r≥10, the **while** loop stops.

1.14   Writing and reading by save and load

save and load
All the data of the user-defined variables currently in the memory space can be saved by

>> save file_name

which saves all the data in the memory into the file named **file_name** with or without an extension. The file created can be found in the current directory. Before or just after you save the data, you may want to see what variables are saved. In this case, use >>**who**, which lists all the variable names that are saved into **file_name**.

The file may be opened in the GUI Editor if you need to see the contents. If you use a name like **file_name.txt** with the extension **.txt**, the file may be opened in *Notepad.* With extension **.doc**, the file may be opened in MS Word.

Let us see more details of what happens. For illustration, we create two variables by

```
>> x=[1 2 3 5]
x =
    1   2   3   5

>> y=[9;4;7;1]
y =
    9
    4
    7
    1
```
Then typing

```
>> save sfile
```

creates a file named **sfile**. Here we assume that no other data than x and y are in the memory, so only x and y should be in **sfile**. The contents of this file opened in GUI Editor are:

```
# Created by Octave 4.0.0, Wed Sep 30 08:09:50 2015 Eastern
Daylight Time <unknown@---->
# name: x
# type: matrix
# rows: 1
# columns: 4
  1 2 3 5


# name: y
# type: matrix
# rows: 4
```

```
# columns: 1
9
4
7
1
```

This file may be loaded any time later by

>>load sfile

Then all the data saved before are recovered in the memory.

Save in ascii and load ascii data file

Data may be saved in ascii format by the **save –ascii** command. For example

>> save –ascii afile x y

saves x and y in ascii format in the **afile** file..

Assuming that the same x and y defined in the prior subsection are saved, the **afile** opened on GUI Editor looks like

```
1.00000000e+000 2.00000000e+000  - - -   5.00000000e+000
9.00000000e+000
4.00000000e+000
7.00000000e+000
1.00000000e+000
```

The data saved in ascii format can be loaded by the load command, but loading a file in ascii format is not quite the inverse of **save** in ascii format. The reason is that while **save -ascii** can save multiple variables, **load** reads the entire file as data of only one variable. Furthermore, the file name becomes the variable's name. For example, if a file named **ydat.tmp** is loaded by

>> load ydat.tmp

the content is loaded to the variable named **ydat** regardless to the extension name.

Therefore, the data file **ydat** must contain data in only one of the following data forms:

(a) a single number
(b) a row array
(c) a column array
(d) a two-dimensional array

If multiple variables have to be loaded, each variable should be saved in a separate ascii data file.

Data files prepared by another computing software such as Fortran or C in ascii (or text) format can be loaded by **load filename.extension** as long as the data structure is one of the four forms mentioned above.

Creating file names automatically

A method to create filenames automatically within an m-file becomes desired some times. If a whole command, including the file-name, is written as a string, it may be executed by **eval**. In the following script, **xdata** is assumed to be computed for each k and saved in separate **files** named **fname001, fname002, ...** in ascii format.

```
for k=1:kmax
%Place here some statements to produce xdata for each k.
%kmax is the maximum number of k (less than 1000)
If k<10,
    s=['save fname00', num2str(k), 'xdata –ascii']
elseif k>=10 & k<100,s=['save fname0', num2str(k)', 'xdata –ascii']
elseif k>=100,
    s=['save fname', num2str(k), 'xdata -ascii']
eval(s)
end
```

1.15　How to write user-defined functions

User-defined functions, which are saved as separate m-files, are equivalent to subroutines and functions in other languages. All the functions, whether it is a built-in function or user-developed function, are written in the form of a function m-file, and they have the same format as follows

function y = function_name(x1, x2, ...)

and saved as function_name.m.

Let us consider developing a function to calculate the equation:

$$y = (2x^3 + 7x^2 - 1)/(x^2 + 5e^{-x})$$

Assuming that the name of the function is **my_function**, the whole m-file for this function is

function y = my_function(x)
y = (2*x.^3 + 7*x.^2 - 1)./(x.^2 + 5*exp(-x));

This function is saved as **my_function.m**. Notice that arithmetic operators are used in case x is an array variable. If x comes as a m-by-n array, the result y is returned as an array of the same size.

If this function is called by

>>z =　my_function(3)

the response is

z = 502.1384

When the argument is an array, for example

x = [3, 1; 0, -11]

then,

>>z = my_function(x)

yields

z =
   1.2542e+001   2.8175e+000
   -2.0000e-001   -6.0636e-003

A user-defined function can return multiple values. For example, the following function is written to return two values, **y** and **z**:

function [y, z]=demo_f(x1, x2)
y = (2*x1.^3 +7*x1.^2 -1)./(x1.^2 + 5*exp(-x2));
z=exp(x1).*sin(x2);

Then, an example of using it is

>>[a1, a2]=demo_f(1.1, 4.2)
a1 =   7.8850
a2 =   -2.6184

Function that calls another function
The argument of a function may be the name of another function. For example, suppose a function that evaluates a weighted average of a function for three different values of argument is given by

$$z = (f(a) + 2f(b) + f(c))/4$$

where $f(x)$ is the function still to be named. The following script illustrates a user-defined function **avf.m** that computes the foregoing equation:

function w = avf(fname, a, b, c)
w = (feval(fname,a)+2*feval(fname,b)
     +feval(fname,c))/4;

In the foregoing script, **fname** is the name of the function in a string variable form. For example, if sin(x) function is used for f(x), with a, b and c pre-defined, we write

>> s = 'sin'
>> z=avf(s, a, b, c)

The z value computed becomes the value of

$$z = (\sin(a) + 2\sin(b) + \sin(c))/4$$

Any function name may be written in place of **sin**.

Debugging of function m-files

Debugging function m-files is more difficult than ordinary script m-files. One reason is that you cannot see the values of variables in function by typing the variable names on the command window. The most basic but effective method of developing a function is to comment out the function statement on the first line by placing **%** before **function** and then test the m-file as an ordinary m-file. Put the function statement back after a thorough examination of the m-file.

Exercise problems for Chapter 1

[1] v=rand(1,100) creates the variable v, which is a row array of 100 random numbers. Make a program to calculate (i) average of the 100 numbers, (ii) determine how many are above 0.5, and (iii) how many are below 0.5.

[2] Calculate the values of $y = x^2 - 2x - 2$ for x=-5, -2, 0, 3, 5 by (i) writing a program using for/end loop, and (ii) not using for/end loop but using the array arithmetic operators.

[3] Write a program to answer Problem [2] such that the function is given by input as a string, and an array of x is given as input.

[4] Write a function that takes x as input, and calculate y=exp(x) if x<0, but $y=1/(1+x^2)$ if x>0.

[5] Write a program that generates N random numbers and calculates how many random numbers fall in the intervals of $0 \leq x < 0.2$, $0.2 \leq x < 0.4$, .., $0.8 \leq x \leq 1.0$, and what fraction of N fell into each group. Calculate the variance of the fractions. Run the program for N=1000, N=10000, and N=100000.

[6] Hilbert matrix is an array of a(i,j) defined by

$$a(i,j)=1/(i+j-1)$$

A hilbert matrix is generated by >>**a=hilb(n)**. Write a program that adds 1 to every entry of 5-by5 hilbert matrix and save it in a file named hilb5.txt. After that, clear the memory by **clear all**, and load the file, hilb5.txt, and print it.

[7] Array variables are defined by

u=
1 2 3 4

v=
1
2
3
4

w=
4 5 6 7

a=
1 4 3
2 -1 1
1 0 2

b=
1 0 1
2 1 1
4 1 2

c=
1 0 0 1
2 1 0 1
3 1 1 0
4 1 2 1

Which of the following operations are valid?

u + v
u + w
c*v
c*w
a+b
b*c
a*b
a.*b
u.^2

[8] Two one-dimensional arrays are defined by

a=[1 2 3]
b=[4 5 6]

Some calculations with a and b became as follows:

```
>> a*b'
ans =   32

>>a'*b
ans =
        4     5     6
        8    10    12
       12    15    18
```

```
>> a.*b
ans =
     4    10    18
>> a.^b
ans =
     1    32    729

>> a.*b'
ans =
     4     8    12
     5    10    15
     6    12    18
```

Explain what happened to each.

[9] A row array of number is defined by d=1:29, that is an array of days in February in a leap year. Make a calendar for February with 7 columns, the first row is Sunday, the second Monday and so on. Assume the first day of February is Monday. The first row may be S, M, T so on, but you may ignore this if it is not easy. The blank days of the calendar may be filled by 0, or blank, which ever you prefer.

[10] An array x has three columns and 10 rows. The first column is numbers starting with 1 and ending at 10, the second column is [0 0.5 1.1 1.6 2.2 2.7 3.4 3.9 4.4 5.0] '. The entries in third column equal those in the second column squared. Print the table as neatly as possible, and print on the top of the table the captions of the columns such as n, x, and x^2.

[11] String variables are defined: A= 'Jones', B= 'Smith', C= 'Alexander', D= 'Xing', E= 'Camden '. Write a program in which all the string variables are first defined. Put all of these string variables into one string variable, Z, where Z is a two- dimensional array of letters, the fist row is Jones, second Smith, and so on.

Chapter 2
Graphics with Octave/Matlab

Graphics plays a central role in Octave/Matlab. Plotting a given data set or the results of computation is possible with very few commands. Octave/Matlab allows you to finish scientific, as well as business, graphics with the highest possible sophistication and elegance.

Trying to understand mathematical equations with graphics is an enjoyable and very efficient way of learning mathematics. Indeed, we might say that *unless you understand mathematical equations graphically, you don't really understand them.* The same applies to scientific and business data. Being able to plot mathematical functions and data freely is the most important step, and this chapter is written to assist you to do that.

For professional people, the importance of graphics is even more profound. Massive data generated today by computers and experiments, as well as from business information source, can only be effectively presented by means of graphic visualization. With Octave/Matlab, it is achieved by plotting data in appropriate forms including one-, two-, and three-dimensional plots, or some times motion pictures (four-dimensional plots).

The best way to learn graphic commands is to read a small amount of instructions at a time and then practice on Octave/Matlab, first by typing examples in the command window or executing the script m-files of this chapter and, second, by changing the m-files in various ways.

2.1 How to Plot

plot
Suppose a set of data points, (x(k), y(k)), k=1, 2,..., n, is to be plotted, where x(k) is an abscissa value and y(k) is an ordinate

value. You need to prepare x and y in an identical array form; namely, x and y are row arrays (or both column arrays) of the same length. Then, the data may be plotted using **plot(x, y)**.

As an example, we plot a function:

$$y = \sin(x) \exp(0.4x), \quad 0 < x < 10$$

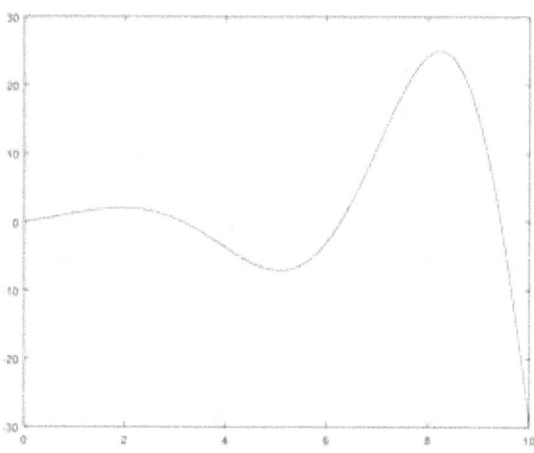

Figure 2.1   A bare bone figure

The following script is a quick attempt:

```
List 2.1
clear, clf, cla
n=201;
delx=10/(n-1);
for k=1:n
   x(k)=(k-1)*delx;
   y(k)=sin(x(k))*exp(0.4*x(k));
end %for
plot(x,y)
```

or more compactly:

```
List 2.1A
clear, clf, cla
```

```
x = 0: 0.05: 10;
y=sin(x).*exp(0.4*x);
plot(x,y)
```

Each of the foregoing scripts plots Figure 2.1.

We assume in the foregoing scripts that the number of data points is 201, then the increment of the x, delx, equals 10/200. Notice in the preceding list, x is a row array of 201 abscissa values of 0, 0.05, 0.1, ..., 10, and y is a row array of ordinate values of the same length. Then, **plot(x,y)** plots the data. The value of n is selected arbitrarily. However, if n is too small, the plotted function loses smoothness.

Figure 2.1 is a bare bone figure because there are no axis labels, no titles, and lines are very thin to see. The line width may be increased by changing **plot(x,y)** to **plot(x,y,'linewidth',4)** where 4 is the line width that could be a smaller or larger value. With this change the figure changes to Figure 2.1A:

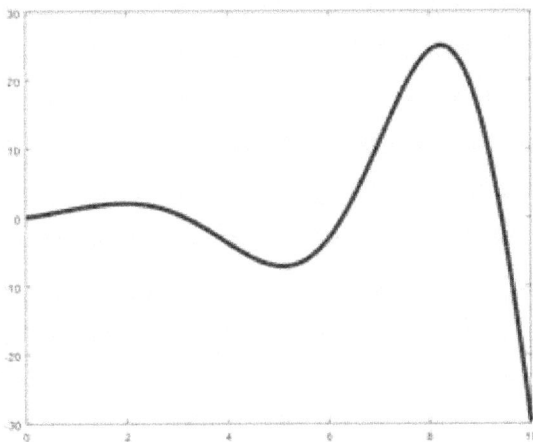

Figure 2.1A   The line width is increased

The default value of the **linewidth** is 0.5. If this is the desired line width, **linewidth** needs not be stated and writing simply **plot(x,y)** suffices.

Further improvements may be done by adding axis labels with a large font size, and making the axis lines thicker. An improved script is List 2.1B, which plots Figure 2.1B.

```
List 2.1B
clear, clf, cla
x = 0: 0.05: 10;
y=sin(x).*exp( 0.4*x);
plot(x,y, 'linewidth',4)
xlabel('x','fontsize',18)
ylabel('y','fontsize',18)
set(gca,'fontsize',16);
set(gca, 'linewidth',2)
```

The axes are labeled by **xlabel('x',fontsize',18)** and **ylabel('y', 'fontsize',18)**. The last two lines in List 2.1B are to make the coordinate lines and tic values thicker, namely the font size of tick values is set to 16 and coordinate line width to 2. These two **set** statements are not necessary in case the default font size, line width of the coordinate lines and tic marks are acceptable. In this book, all figures are reduced in size, so without the increase in **linewidth** and **fontsize**, the figures would be too thin to read.

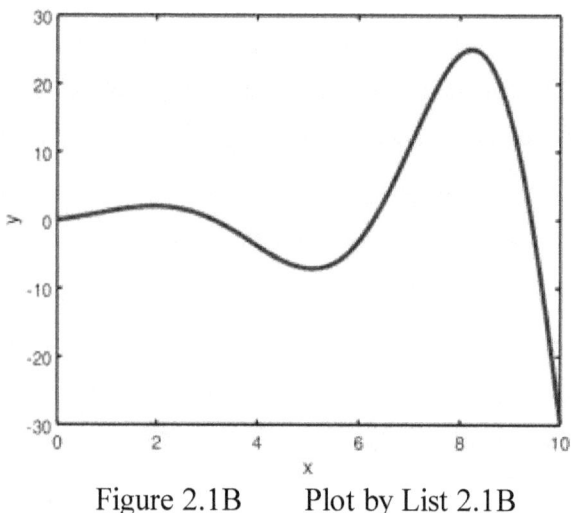

Figure 2.1B   Plot by List 2.1B

Figure 2.2 is plotted by List 2.2 connecting a series of points in a complex plane.

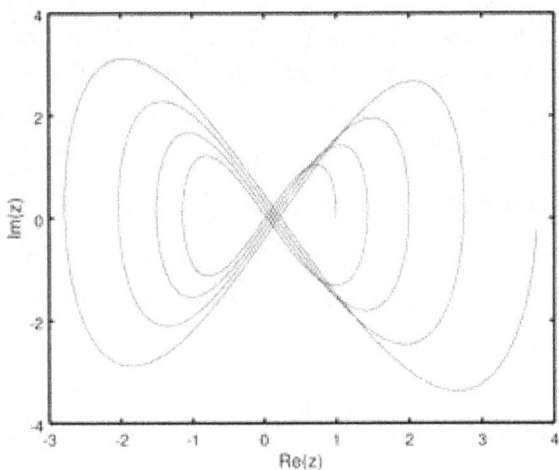

Figure 2.2   Plot on a complex plane using List 2.2

```
List 2.2
clear, clf, cla
p=0:0.05:8*pi;
z=(cos(p) + i*sin(2*p)).*exp(0.05*p) + 0.01*p;
plot(real(z), imag(z))
xlabel('Re(z)','fontsize',16)
ylabel('Im(z)','fontsize',16)
```

Plotting by marks only

Data can be plotted by marks only without connecting by lines.
Nine samples of marks are illustrated here:

Point :   .
Plus :   +
Star :   *
Diamond: d
Circle :   o
Pentagon : p
Square :   s
Triangle :   ^
x-mark :   x

To find more marks, type **help plot**. To plot with one type of mark only, place the mark symbol as a string after the coordinates in the arguments of **plot**. The graph produced by List 2.3 is shown in Figure 2.3.

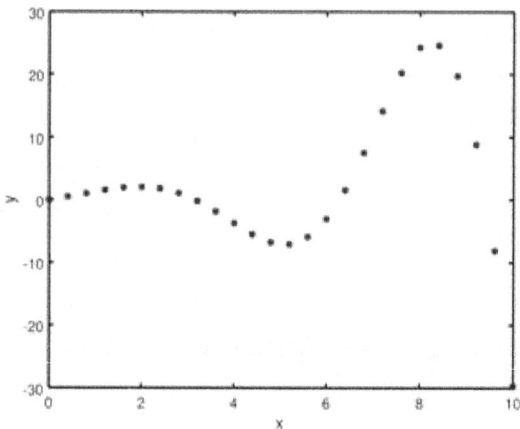

Figure 2.3   A graph plotted with marks only

```
List 2.3
Clear, clf, cla
x = (0:0.4:10);
y=sin(x).*exp(-0.4*x);
plot(x,y,'*', 'linewidth', 4);
xlabel('x','fontsize',16)
ylabel('y','fontsize',16)
set(gca,'fontsize',14);
set(gca, 'linewidth',2)
```

Line types
Four line types are available as shown next:

Line Type Symbols

Line type	Symbol
Solid	-
Dash	--
Dotted	:
Dashdot	-.

Default is the solid line type. To plot with a selected line type, specify the line mark after x and y, for example:

plot(x,y,'-.')

If thick lines are desired, add **'linewidth', n** before the closing the right parenthesis, where n is a line width number such as 2, 4 or 6, for example:

plot(x,y,'-.','linewidth', 4)

Plotting a function with both lines and marks

Plot twice, namely, the first time by the default solid line and the second time with marks only. To plot with a solid line and a mark +, for example, write the plot command as **plot(x,y,x,y,'+')**, where '+' applies to the second pair of **x** and **y**. The **text** command may be used to plot with any mark or letter; however, the location of the mark may be slightly offset from the correct location of the data point.

Line colors

Eight colors, namely, red, yellow, magenta, cyan, green, blue, white, and black, are available for the lines and marks. These colors are specified by letters, r, y, m, c, g, b, w, and k, respectively. Use the color symbol just like the line types in the argument of plot, for example:

plot(x, y, 'g')

A combination of mark and color is also possible:

plot(x, y, '+g')

plots the data with + marks in green.

Cleaning and clearing graph

clf clears everything inside the graphic window, while **cla** clears the plotted curves and redraws the axes.

Figure's number

The command **figure** opens a new graphic window that is numbered consecutively from the previous one, while **figure(n)**, where n is an integer, opens the figure window numbered by n. If multiple figure windows exist, you have to be aware which one is the current figure. This is because all the graphic commands apply to the current figure. The latest window opened is the current window unless an older one is called for by **figure(n)**. The sequential number n is displayed at the top left corner of the figure window.

Figure's position on the computer monitor

The size and shape of figure on the computer monitor are determined by default. However, the size, shape and location on the computer monitor may be changed by

figure(n,'position', [pix,piy,pwx,pwy])

where **pix** and **piy** are the horizontal and vertical pixel coordinates, respectively, of the left bottom corner of the figure in the monitor pixel coordinates (the origin of the pixel coordinates is at the left bottom corner of the monitor); **pwx** is the number of pixels in the width; and **pwy** is the number of pixels in the height of the figure window. By specifying **[pix,piy, pwx, pwy]** appropriately, a desired location on the monitor screen, size and shape may be achieved. The existing figure may also be changed by **set(gcf, 'position', [pix, piy, pwx, pwy])**. The coordinate values of the current figure may be obtained by **get(gcf, 'position')**.

With the **figure** command, it is also possible to display multiple figures in neatly organized manner on the monitor.

close

close(n) closes **figure(n)**, and **close all** closes all figures.

axis, axis on, axis off

For a figure, the minimum and maximum of the coordinates, tic marks, and the coordinate values at the tic marks, are all determined automatically. Some properties of a figure may be changed as illustrated next.

The coordinate axes and tic marks can be removed by

axis off

which may be written in an m-files, or typed on a keyboard while the figure is open. The axes and tics are reinstated by **axis on**.

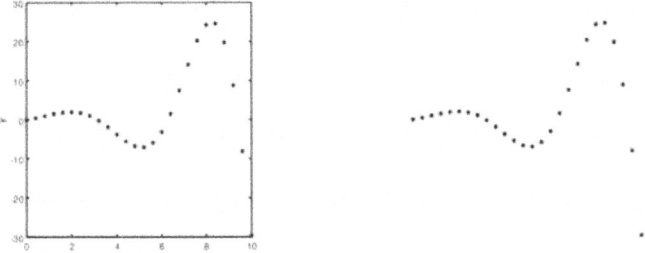

Figure 2.4   A plot with **axis square** (left), and a plot with **axis off** (right)

The maximum and minimum of the coordinates of the graph may be specified by

axis([x-min, x_max, y_min, y_max])

Any lines outside the limits will be clipped. This command can be used in an m-files, but also can be typed on the keyboard so that the view area can be changed as many times as desired while the figure is on the monitor screen. It is suggested that the reader appends **axis([-10, 20, -20, 30]),** as a test trial, to List 2.3 to see the effect of axis.

Grid on, grid off

A grid can be added to the graph by **grid on**, while **grid off** removes the grid. Simply using **grid** multiple times toggles **grid on** and **off**. An example of using **grid on** is illustrated in Figure 2.5 plotted by List 2.4.

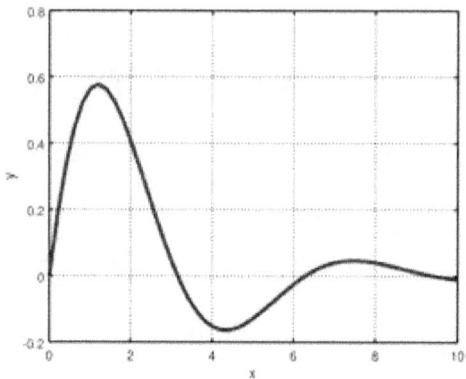

Figure 2.5   A figure with `grid on`

List 2.4
```
clear,clf,cla
x = (0:0.2:10)';
y=sin(x).*exp(-0.4*x);
plot(x,y,'linewidth', 4)
grid on
xlabel('x','fontsize',16); ylabel('y','fontsize',16)
set(gca,'fontsize',14); set(gca,'linewidth',2)
```

Polar plot

A function on a polar coordinate can be plotted by **polar**. Figure 2.6 is plotted by List 2.5.

List 2.5
```
clear,clf,cla
t = 0:.05:pi+.01;
y = sin(3*t).*exp(0.3*t);
polar(t,y)
title('Polar plot')
grid on
set(gca,'fontsize',14);
set(gca, 'linewidth',2);
```

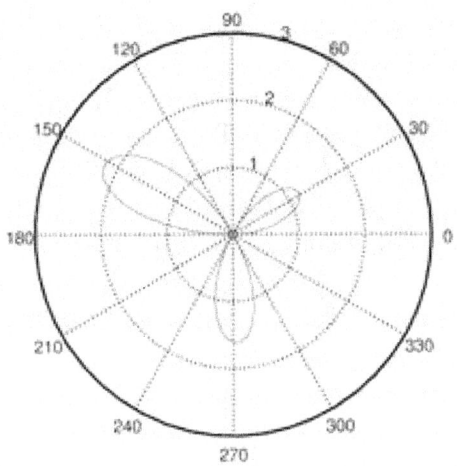

Figure 2.6   Polar plot

Log and semi-log plot

A function may be plotted on a log-log scale by **loglog**. See List
2.6 and Figure 2.7.

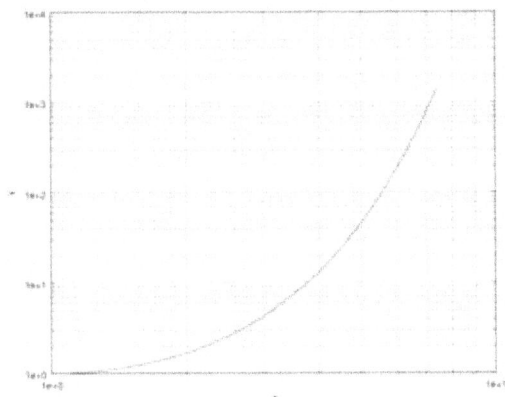

Figure 2.7   A log-log plot

List 2.6
```
clear,clf,cla
t = .1:.1:2;
x = exp(t);
y = exp(t.*sinh(t));
loglog(x,y)
```

```
grid on
xlabel('x');
ylabel('y')
```

In this m-file, **xlabel** and **ylabel** are used without fontsize, but with the default font size. Also **set(gca ...** is not used. These are intentional for the purpose of illustrating the effect of using the default font size and default line width for the coordinate lines.

A semi-log plot with the log scale for y is produced by List 2.7. See Figure 2.8.

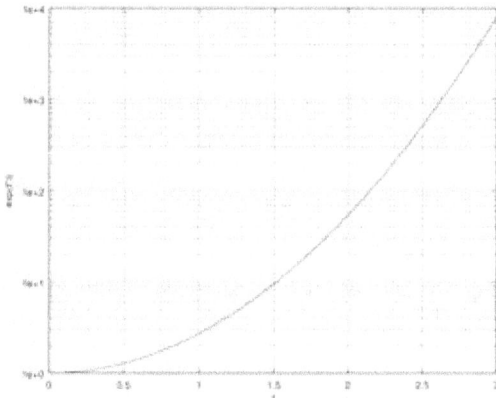

Figure 2.8   A semi-log plot

```
List 2.7
clear,clf,cla
t = .1:.1:3;
semilogy(t, exp(t.*t))
grid on
xlabel('t'); ylabel('exp(t*t)')
```

Plotting an implicit function

If a function is given implicitly, for example,

$$y^3 + \exp(y) = \tanh(x)$$

it cannot be expressed by x as a function of y, nor y as a function of x.

70

The function can be plotted, however, using **contour**. More detail of this approach is discussed in the subsection for contour of Section 2.3.

Plotting multiple curves

To plot two or more curves with a single **plot** command, write all pairs of coordinates repeatedly in the **plot** command:

```
List 2.8
clear, clf, cla
x = 0:0.05:5;
y = sin(x);
z = cos(x);
plot(x,y,x,z)
```

Different line type or color is automatically selected for each curve by default. If desired, however, selected line color, line type, or mark, may be specified after each pair of coordinates; for example,

plot(x, y,'-g', x, z,':')

Octave allows to place '**linewidth**' for each data set to be plotted like

plot(x, y,'-g', 'linewidth', 1, x, z,':r', 'linewidth', 4)

On the other hand, Matlab despises the **linewidth** before the second data set. So plot each data set separately as

plot(x, y,'-g', 'linewidth', 1); hold on
plot(x, z,':r', 'linewidth', 4)

hold on, hold off

Until now we plotted all the curves at once with a single **plot** command. It often becomes desirable, however, to add a curve to a graph that has already been plotted. Such additional plotting can be done using **hold on** (see Figure 2.9 plotted by List 2.9).

```
List 2.9
clear,clf,cla
x = 0:0.05:5;
y = sin(x); plot(x,y); hold on
z = cos(x); plot(x,z,'--')
xlabel('x', 'fontsize',14); hold off
ylabel('y(-) and z(--)','fontsize',14)
```

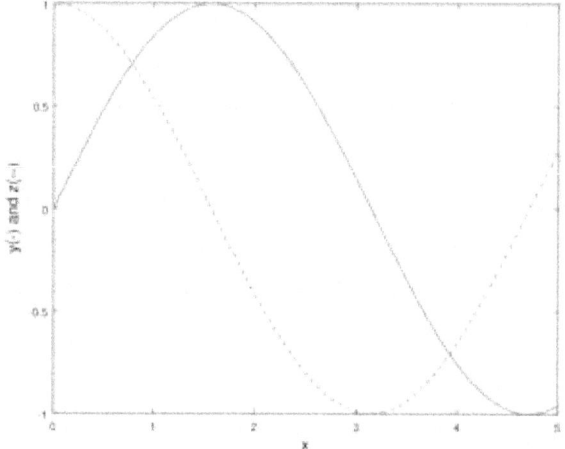

Figure 2.9   Two curves plotted with **hold on**

Once the **hold on** command is issued, the graph may stay on even when another script is executed. Therefore, it is prudent to place a **hold off** command at the end of the script.

When multiple curves are plotted with **hold on**, it is desirable to specify minimums and maximums of the coordinates on the graphic domain by the **axis** command. Otherwise, the limits are determined by default based on the first curve, which may cause other curves to be clipped.

The **hold on** command also becomes important when a time-consuming plot is undertaken for the following reason. The command to change parameters for figures such as **axis**, **colormap**, **view**, and other parameters can be used after a figure is plotted, but each time a new command is issued, the whole figure is re-plotted. In order to save time, give all the parameter commands before plotting, hold with **hold on**, and then use **plot**.

legend
The legend in a figure can be added by the **legend** command. The legend in Figure 2.10 is created by **legend('sin(x)', 'cos(x)')** added in List 2.10:

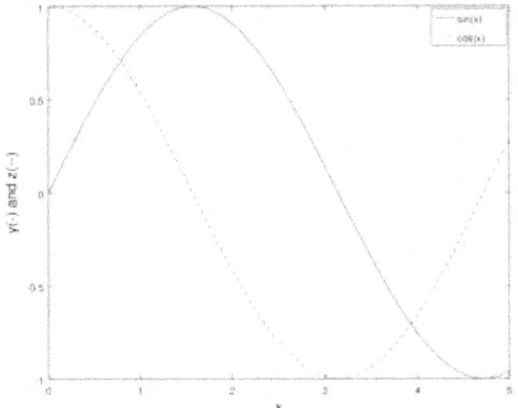

Figure 2.10   Illustration of the use of the
legend command

List 2.10
clear,clf,cla
x = 0:0.05:5;
y = sin(x); plot(x,y); hold on
z = cos(x); plot(x,z,'--')
xlabel('x', 'fontsize',14);
ylabel('y(-) and z(--)','fontsize',14)
legend('sin(x)','cos(x)')

title
A title may be added to the top of the figure by the **title** command.

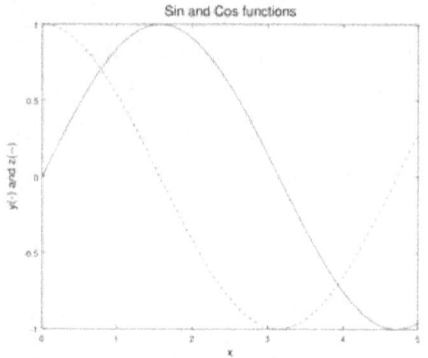

Figure 2.11   Figure with a title on the top

List 2.11
clear,clf,cla
x = 0:0.05:5;
y = sin(x); plot(x,y); hold on
z = cos(x); plot(x,z,'--')
xlabel('x', 'fontsize',14);
ylabel('y(-) and z(--)','fontsize',14)
title('sine and cosine functions','fontsize',16)

text

Text can be written in a graph by **text**. It needs three parameters in the argument, namely, **text(x, y, 'string')**. The first two are x and y values of the location where the string starts. The third is a string variable to be printed. The string variable can be a text enclosed by quote signs or a predefined string variable. For illustration, two **text** commands are used in List 2.12 to plot Figure 2.12.

List 2.12
clear,clf,cla
x = 0:0.05:5;
y = sin(x); plot(x,y); hold on
z = cos(x); plot(x,z,'--')
xlabel('x', 'fontsize',14);
ylabel('y(-) and z(--)','fontsize',14)
text(1.65, 0, 'sin(x)' ,'fontsize',16)
text(3.2, 0, 'cos(x)' ,'fontsize',16)

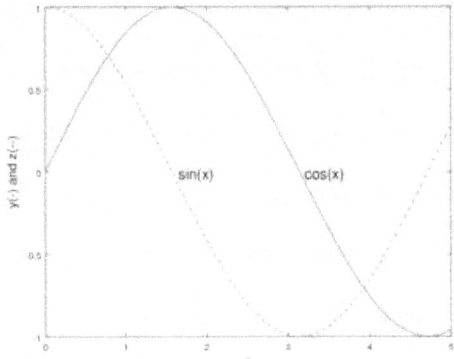

Figure 2.12   Illustration of the use of **text**

Text colors and other properties of text

Color and font size of a text in the graph may be changed. For example,

text(0.3, 0.2 ,'string','fontsize',18,'color','r')

will print string in red color with font size 18. If a default color is to be changed to green, for example, use:

set(gcf,'DefaultTextColor', 'g')

Thereafter, the text will be typed in green unless specified in each **text** command. Color for text may be chosen from red, yellow, green, cyan, blue, and magenta, which are abbreviated by 'r', 'y', 'g', 'c', 'b', and 'm', respectively. Color may be changed as many times as necessary.

Greek symbols may be printed by **text**. For example,

Figure
axis([1.4 1.8   -0.1 0.1])
text(1.5,0, '\alpha\beta\delta\epsilon\gamma\Gamma
\Omega', 'fontsize',36)

will print Greek letters in a figure, as illustrated below:

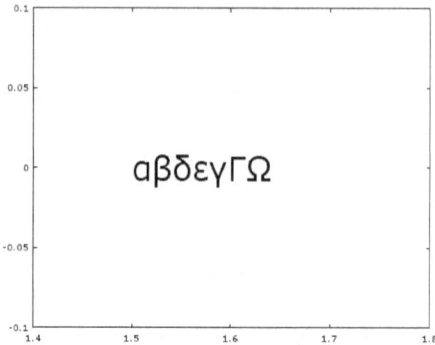

Other symbols may also be included, all according to the LaTex.

The font size of the axis tic mark values may be changed by the **set** command; for example,

set(gca, 'fontsize',18)

changes the font of axis to size 18.

Superscript and subscripts may be written in a graph by **text**. To print a single character or a group of characters as superscript, use the caret symbol ^ followed by a single letter or group of letters enclosed by {}. For example, **text(2, 4, ' x^{-2} ')** will print x^{-2} as text at x=2 and y=4. To write a subscript is the same as superscript except the ^ is replaced by underscore: for example, text(2, 4, ' s_{i,j}') will print $s_{i,j}$ as text. For illustration,

```
Figure
axis([1.4 1.8   -0.1 0.1])
text(1.5,0, 'y_{i,j} =   x^{2}', 'fontsize',36)
```

plots the following figure:

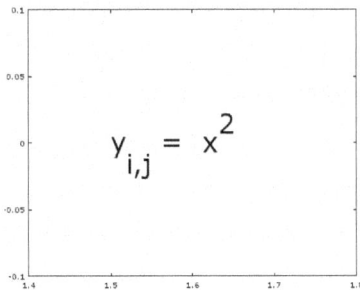

subplot

With **subplot(m,n,k)** command, mxn graphs are plotted in a single figure, where m, n, and k are integers. Here, the pair of m and n means a mxn array of graphs, and k is the sequential number of the graph. For example, **plot** after **subplot(3,2,1)** will plot the first graph in the 3x2 figures. The following script plots four graphs, as illustrated in Figure 2.13:

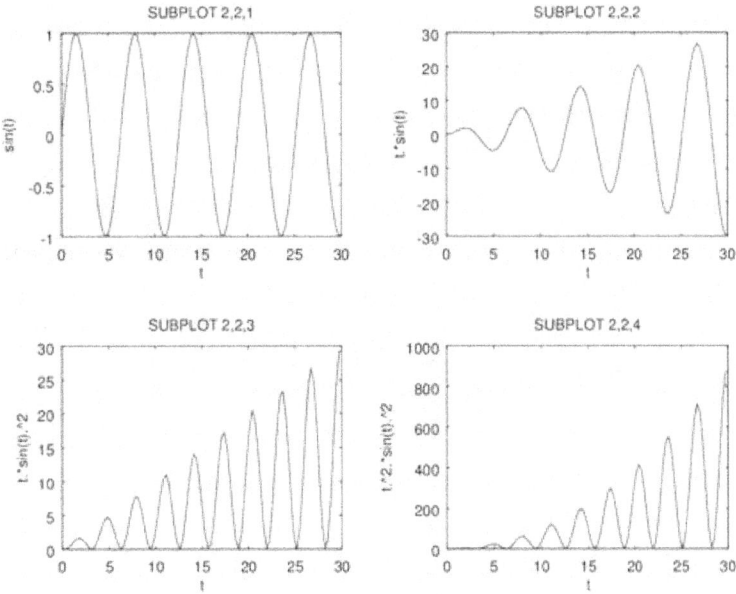

Figure 2.13   Plotting multiple figures by **subplot**

List 2.13
clear,clf,hold off

```
t=0:.3:30;
subplot(2,2,1),
   plot(t,sin(t)),title('SUBPLOT 2,2,1')
   xlabel('t'); ylabel('sin(t)')
subplot(2,2,2),
   plot(t,t.*sin(t)),title('SUBPLOT 2,2,2')
   xlabel('t'); ylabel('t.*sin(t)')
subplot(2,2,3),
   plot(t,t.*sin(t).^2),title('SUBPLOT 2,2,3')
   xlabel('t'); ylabel('t.*sin(t).^2')
subplot(2,2,4),
   plot(t,t.^2 .*sin(t).^2),title('SUBPLOT 2,2,4')
   xlabel('t'); ylabel('t.^2.*sin(t).^2')
```

A vertical stack of two graphs is plotted by

```
subplot(2,1,1), plot(
subplot(2,1,2), plot(
```

Likewise, a row of two graphs is plotted by

```
subplot(1,2,1), plot(
subplot(1,2,2), plot(
```

3d plot

The command **plot3** is the three-dimensional version of **plot**. All the rules explained for **plot** apply to **plot3**. A spiral motion of a particle from point A to B in Figure 2.14 is plotted by List 2.14. The view angle may be changed by **view**, as explained in more detail in Section 2.3. The axis command,

```
axis([x-min,x-max,y-min,y-max,z-min,z-max])
```

may be used to define bounds of the three-dimensional space.

```
List 2.14
clear,clf
t=0:0.1:20;
r= exp(-0.2*t);
th=pi*t*0.5;
```

```
z=t;
x=r.*cos(th);
y=r.*sin(th);
plot3(x,y,z, 'linewidth', 3)
hold on
plot3([1,1], [-0.5,0], [0,0], 'linewidth', 3)
text( 0.9,-0.95,1, 'A', 'fontsize',14)
n=length(x);
text( x(n),y(n),z(n)+2,'B', 'fontsize',14)
xlabel('X', 'fontsize',14);
ylabel('Y', 'fontsize',14);
zlabel('Z', 'fontsize',14);
```

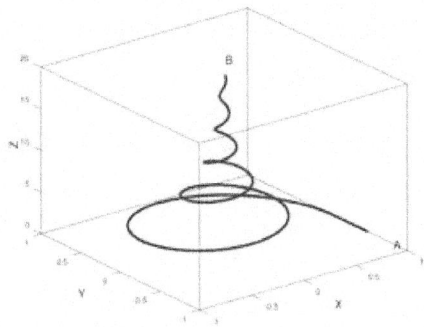

Figure 2.14   3D plot with `axis`

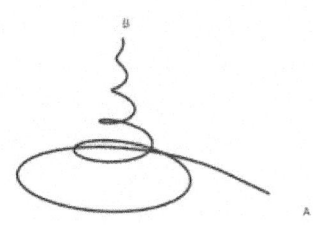

Figure 2.15   3D plot with `axis off`

2.2 How to print or record graphs

print

The command to produce a graphic file of the current figure in jpeg form is **print filename.jpg** or **print -djpg filename**. To create a post script file of a figure, the command is **print filename.ps** or **print -dps filename**. See the next subsection in case the figure includes texts with subscripts or superscripts or Greek letters.

The figures in jpeg form may be inserted in the MS Word. The figure file in the post script form may used in LaTex editor.

Copy & paste

Figures can be copied from figure windows and pasted on MS Word by copy & paste operation using mouse.

Unfortunately, the figures that include texts with subscripts/superscripts or Greek letters may not come out well with the **print** command or by copy & past. In this case, a screen capture technique is recommended. That is, click on the figure to copy. Push *Ctrl key* and *Alt key* and *Print Screen key* together. This action copies the entire figure window including its frame. Open the *Paint* software on PC. Paste on it by pushing *Cntl key* and *character v key* together. In the Paint software, you will see the figure, but it includes the window frame. The Paint software has a rubber band tool by which you can select the area of the figure excluding the figure window frame. Copy the selected area, and paste on MS Word. If you save the Word document in html format with a **filename**, the figure in jpg format becomes available in the file named **filename_files**.

Plotting by black only

There is a little secret in creating the 3-dimensional plots in this book. Because they are plotted in the figure window with color, some colors, particularly the colors near yellow, fade out because the book is printed by black only. Therefore it became necessary to plot only in black. This was done by typing the command

```
set(gcf, 'colormap', zeros(64,3))
```

on the keyboard after a color graph is plotted. This command changes the color map to all black whatever the original color is.

2.3   Plot of two-dimensional functions

Mesh plot

Suppose $x(i)$, $i=1,2,$ ... imax, are values on the x coordinate in increasing order, and $y(j)$, $j=1,2,$ ...jmax, the values on the y coordinate in increasing order. Consider vertical lines passing through the points on the x-coordinate, namely, $x = x(i)$, and horizontal lines passing through the points on the y-coordinate, namely, $y = y(j)$, then the intersections between the two families of the lines form a mesh. A mesh point in this mesh may be denoted by the pair of indices (i,j) or pair of coordinates $(x(i), y(j))$. If we define a two-dimensional array of x as

$$x(i,j) = x(i)$$

and that of y as

$$y(i,j) = y(j)$$

then a function on the two-dimensional coordinate, $z = f(x,y)$, may be discretely represented by an array z:

$$z(i,j) = f(x(i,j), y(i,j))$$

In writing an m-file script, we denote the one-dimensional array of $x(i)$ by x1, and the same of $y(j)$ by y1; the two-dimensional array of $x(i, j)$ by x2, and the same of $y(i, j)$ by y2. Once the values of x1 and y1 are set, then two-dimensional arrays, x2 and y2, can be generated by the **meshgrid** command as

$$[x2, y2] = meshgrid(x1, y1)$$

We assume the function $z = f(x, y)$ is given by

81

$$z = f(x,y) = x \exp(-x^2 - y^2), \qquad -2 < x < 2, \qquad -2 < y < 2$$

then, the two-dimensional array z is computed by

$$z = x.*\exp(-x.^2 - y.^2)$$

A three-dimensional plot of z is done by the **mesh** command as

$$\text{mesh}(x, y, z)$$

A whole script is shown in List 2.15, which produces Figure 2.16.

```
List 2.15
clear, clf
x1 = -2:.2:2;
y1 = -2:.2:2;
[x2,y2] = meshgrid(x1,y1);
z2 = x2 .* exp(-x2.^2 - y2.^2);
mesh(x2,y2,z2)
title('This is a 3-D plot of   z = x * exp(-x^2 - y^2)')
xlabel('x'); ylabel('y'); zlabel('z');
```

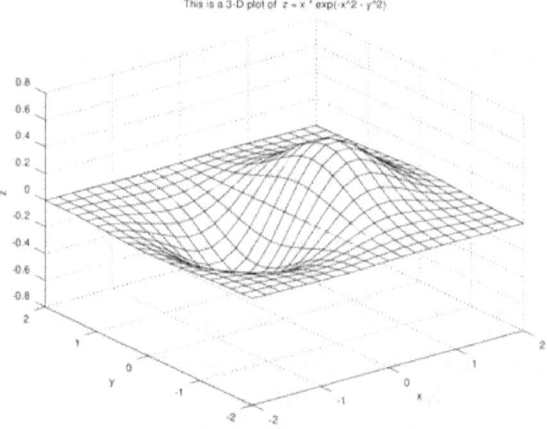

Figure 2.16   3D plot by **mesh**

In Figure 2.16, the three-dimensional plot is viewed from default view eyes in space, but the location of the view eyes may be changed by the **view** command. In Figure 2.17, the view point is changed to [0.1, 1, 0.5], or equivalently x=0.1, y=1, and z=0.5.

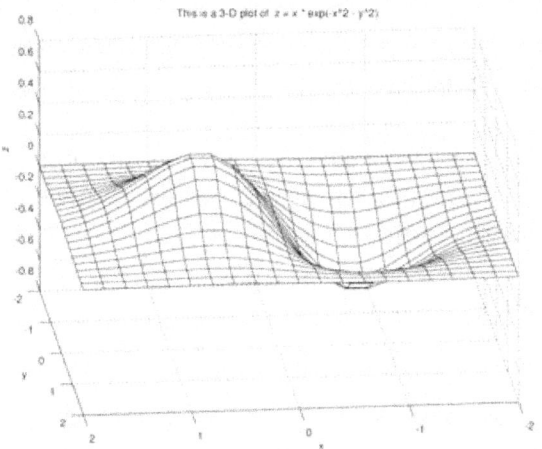

Figure 2.17   3D plot with a changed **view**

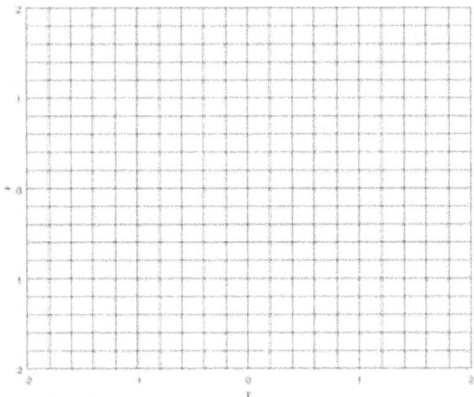

Figure 2.18   Plot of the mesh on the x-y plane

Figure 2.18 is a plane view of the mesh on the x-y plane plotted by **mesh**:

```
mesh(x2,y2,0*x2);
view([0,0,100001]);
xlabel('x'); ylabel('y')
```

Contour plot

Contour is another way of visually expressing the distribution of a two-dimensional function.

Assuming that the mesh grid data and functional data used in the prior section are utilized again, we simply write them in the command to plot contour:

contour(x, y, z, level)

where **level** is an array of contour levels, which are based on the interval between minimum and maximum values of z divided into a given number of subintervals. The number of levels is 10 in the present example. Contour levels are the mid points of the subintervals. Figure 2.19 of contour is plotted by List 2.16.

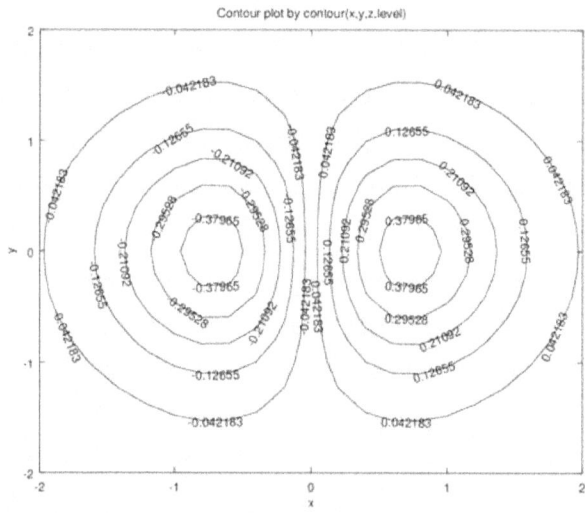

Figure 2.19   Contour plot of $z = x \exp(-x^2 - y^2)$

List 2.16
```
clear, clf
x1 = -2:.2:2;
y1 = -2:.2:2;
[x2,y2] = meshgrid(x1,y1);
z2 = x2 .* exp(-x2.^2 - y2.^2);
zmax=max(max(z2)); zmin=min(min(z2));
dz = (zmax-zmin)/10;
level = zmin + 0.5*dz: dz: zmax;
h=contour(x2,y2,z2,level); clabel(h)
title('Contour plot by contour(x,y,z,level)')
xlabel('x'); ylabel('y');
```

Plotting an implicit function using contour

Plotting of an implicit function mentioned earlier in Section 2.1

$$y^3 + \exp(y) = \tanh(x)$$

is now possible by **contour**. We rewrite the foregoing equation and define a two-dimensional function $z(x, y)$ as

$$z(x, y) = y^3 + \exp(y) - \tanh(x)$$

We plot its contour for only one level of z=0. The plot of this function is illustrated in Figure 2.20, plotted by the script in List 2.17.

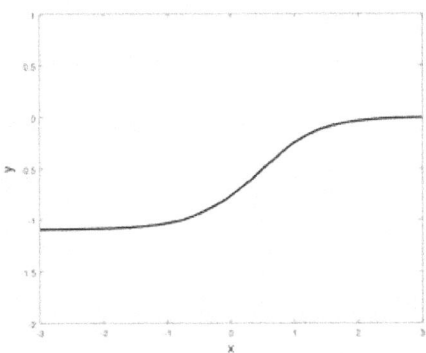

Figure 2.20   Plot of $z = y^3 + \exp(y) - \tanh(x)$

85

List 2.17
```
clear, clf, cla
xm = -3:0.2:3;   ym = -2:0.2:1;
[x, y] = meshgrid(xm, ym);
f = y.^3 + exp(y) - tanh(x);
contour(x,y,f,[0,0],'linewidth',2)
xlabel('x', 'fontsize',16);
ylabel('y', 'fontsize',16)
```

quiver

The **quiver** command is designed to plot 2D fluid velocity distributions and plots a vector indicating the magnitude and direction of the flow at each grid point. The quiver command requires the 2-dimensional arrays of x and y, plus u and v where u is the velocity component in the x-direction and v the velocity component in the y-direction. The format of this command is **quiver(x,y,u,v)**.

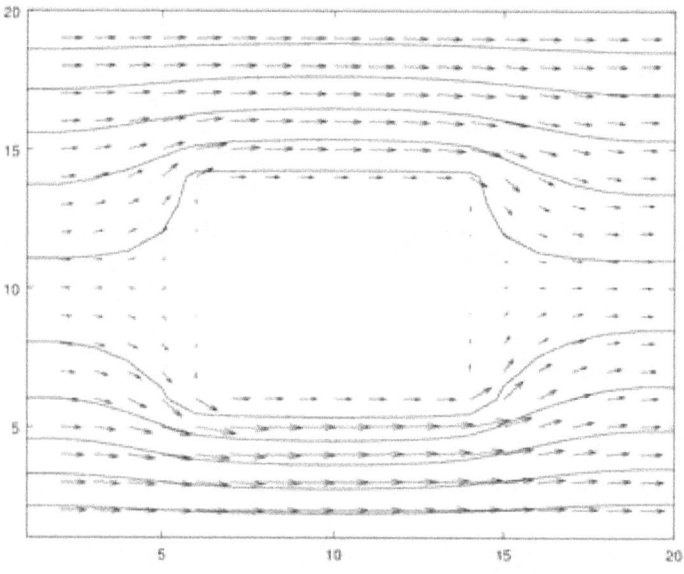

Figure 2.21 Velocity vector plot with stream function contour

In List 2.18, Part A solves the stream function equation for the inviscid flow, Part B computes the velocity distributions u and v, and in Part C, the velocity vectors are plotted by **quiver**, and the stream function is plotted by **contour**.

The results of plot are shown in Figure 2.21.

```
List 2.18
clear
clf
% PART A
imax=20;
jmax=20;
for i=1:imax
    for j=1:jmax
        x(i,j)=i;
        y(i,j)=j;
        f(i,j)=1;
        if j==1,    f(i,j)=jmax; end
        if j==jmax,    f(i,j)=1; end
    end
end
f(1:2,1:2)

i1=imax/2-5
i2=imax/2+5
j1=jmax/2-5
j2=jmax/2+5

for k=1:600
    for i=1:imax
        for j=2:jmax-1
            if i==1, f(i,j)=f(i+1,j); end %if
            if (i>1&i<imax),
                f(i,j)=0.25*( f(i-1,j)+f(i+1,j) + f(i,j-1)+    ...
                    f(i,j+1))*1.9-0.9*f(i,j);
                end %if
            if (i==imax) f(i,j)=f(i-1,j);
            end %if
            if (i>i1 & i<i2 & j>j1 & j<j2),f(i,j)=jmax/2;
            end %if
        end
    end
end

% PART B
```

87

```
u(imax,jmax)=0;
v(imax,jmax)=0;
for i=2:imax-1
    for j=2:jmax-1
        u(i,j) = -(f(i,j+1)-f(i,j-1))/2;
        v(i,j) = (f(i+1,j)-f(i-1,j))/2;
    end
end

%PART C
quiver(x,y,u,v), hold on
contour(x,y,f)
```

2.4 Plotting of 3-dimensional structures

Octave/Matlab have several graphic commands to construct 3D images. Some examples are illustrated here. Scripts used to plot the following figures are listed in http://octave.ismr.us/Commuter-Airplane.htm

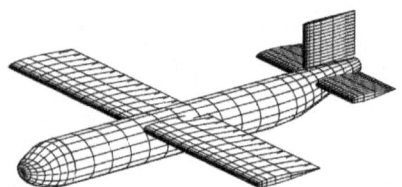

Plot 1: Plotted by **mesh** command

Plot 2: Plotted with **surfl** and **shading interp**

2.5 Bar charts

The graphs in this section are in color when produced by Octave/ Matlab, which needed to be printed only in black/white in the book. However, the same graphs in color can be seen in http://octave.ismr.us/bar-color.htm.

With the bar(x) command, a bar chart is drawn, where x is an array of the data.

>> bar([1 3 2 1])

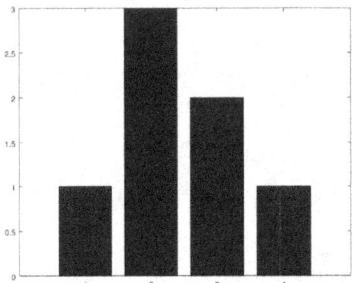

If x is a two-dimensional array of two rows, each column is considered as a category, and each row is considered to be a group such as a year or month.

>> bar([1 3 2 1; 2 1 1 3])

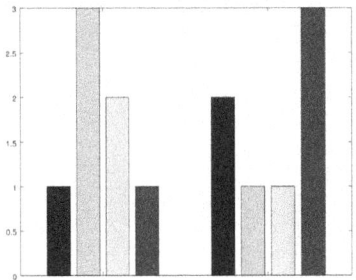

With the bar(x, "stacked"), the bars in each group are stacked.

>> bar([1 3 2 1; 2 1 1 3], "stacked")

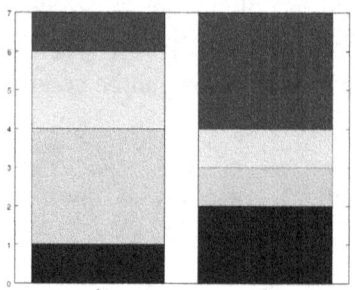

The width of bars are changed by w in bar(x,w,...). In the following example w=0.3 is used:

>> bar([1 3 2 1; 2 1 1 3], 0.3,"stacked")

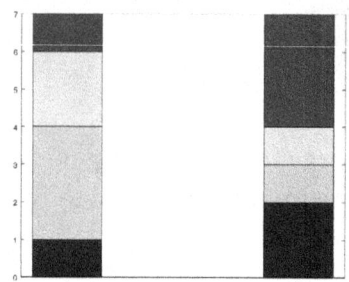

The bar(x, w, "hist") creates a bar chart in the histogram style:

>> bar([1 3 2 1 6 4; 2 1 1 0.5 3 5], 0.8,"hist")

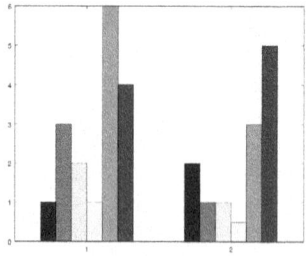

The bar(x, w, "histc") is the same as bar(x, w, "hist") except the histograms is left adjusted:

>> bar([1 3 2 1 6 4; 2 1 1 0.5 3 5], 0.8,"histc")

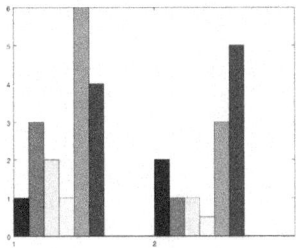

Legends may be added by the **legend** command:

>>bar([1 3 2 1 6 4; 2 1 1 0.5 3 5], 0.8,"histc")
>>axis off; legend('a', 'b', 'c', 'd', 'e', 'f')

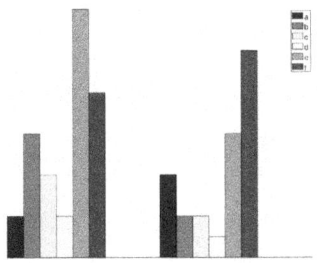

By removing the axis and adding texts, the meaning of each group can be printed under each group:

```
>>clf
>>bar([1 3 2 1 6 4; 2 1 1 0.5 3 5], 0.8,"histc")
>>axis([1 3 -0.5 6]); axis off;
>>legend('a', 'b', 'c', 'd', 'e', 'f')
>>text(1.2,-0.3, 'Year 1', 'fontsize', 16)
>>text(2.2,-0.3, 'Year 2', 'fontsize', 16)
```

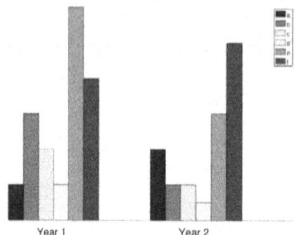

Year 1   Year 2

Exercise problems for Chapter 2

[1] Plot the following functions in -1≤x≤1 using the color specified for each. The graph needs xlabel and ylabel in legible font size. The curves must be plotted with linewidth 2.

$$y=\sin(x)\exp(-x) \qquad \text{(green)}$$
$$y=\cos(5a\cos(x)) \qquad \text{(blue)}$$

[2] Plot the following functions together using the method of plotting implicit functions for -1<x<3, -2<y<3.

$$2(x - 1)^2 + (y - 1)^2 = 3$$
$$(x^{1.5} - 3)^2 + y^2 = 2$$

[3] Plot the following function with xlabel and ylabel:

$$y = 1/(1 + (x-2)^2), \quad 0 \leq x \leq 4$$

[4] Plot y=tan(x) in 0 ≤ x ≤ 10. Does it look awesome or awkward? Propose how to make it awesome.

[5] Plot the following equation in x>0 and y>0.

$$\sqrt{(1+x^2+\exp(y))} = 4\sin(xy)$$

[6] Repeat [1] and write the function near the curve of each as text.

[7] Repeat [1] and add the definition of the function as legends.

[8] Plot contour graph of the following function in x>0 and y>0 with contour lebels.

$$f(x,y) = 1/(\ 1+ (x-1)^2 + (y-1)^2\ )$$

[9] Plot the following data in a bar graph

Year	Quantity a	b
1	25	12
2	31	13
3	29	17
4	40	19
5	35	21

Chapter 3
Numerical Methods

This chapter briefly introduces numerical methods with implementation in Octave/Matlab but without theoretical details. For
fundamentals and theoretical details, the readers are recommended
to read another book of the author, *Foundation of Numerical
Analysis*.

3.1 Matrix and vectors

A matrix is a rectangular array of numbers. A column vector is a
column array of numbers, and a row vector is a row array of
numbers in Octave/Matlab. Arrays variables in Octave/Matlab are
used as matrices and vectors in linear algebra.

Addition and subtraction of matrices and vectors follow the
same rules for those of the array variables. Multiplication of
vectors and matrices is, however, different from the arithmetic
multiplication of array variables. Let us start with two vectors
defined by

$$u = \begin{pmatrix} u_1 & u_2 \end{pmatrix}$$

$$v = \begin{pmatrix} v_1 \\ v_2 \end{pmatrix}$$

(3.1)

where u is a row vector, and v is a column vector. The product uv
becomes

$$uv = \begin{pmatrix} u_1 & u_2 \end{pmatrix} \begin{pmatrix} v_1 \\ v_2 \end{pmatrix} = \begin{pmatrix} u_1 v_1 + u_2 v_2 \end{pmatrix}$$

(3.2)

In Octave/Matlab, the vector product is written as $u*v$. If the order
of u and v are changed to vu, the result is

$$vu = \begin{pmatrix} v_1 \\ v_2 \end{pmatrix} (u_1 \quad u_2) = \begin{pmatrix} v_1 u_1 & v_1 u_2 \\ v_2 u_1 & v_2 u_2 \end{pmatrix} \tag{3.3}$$

Now we consider a matrix times a vector. Define a matrix by

$$a = \begin{pmatrix} a_{1,1} & a_{1,2} \\ a_{2,1} & a_{2,2} \end{pmatrix} \tag{3.4}$$

The product of a and u becomes a column vector:

$$au = \begin{pmatrix} a_{1,1}u_1 + a_{1,2}u_2 \\ a_{2,1}u_1 + a_{2,2}u_2 \end{pmatrix} \tag{3.5}$$

Let us define another matrix b by

$$b = \begin{pmatrix} b_{1,1} & b_{1,2} \\ b_{2,1} & b_{2,2} \end{pmatrix} \tag{3.6}$$

The product a and b becomes

$$ab = \begin{pmatrix} a_{1,1}b_{1,1} + a_{1,2}b_{2,1} & a_{1,1}b_{1,2} + a_{1,2}b_{2,2} \\ a_{2,1}b_{1,1} + a_{2,2}b_{2,1} & a_{2,1}b_{1,2} + a_{2,2}b_{2,2} \end{pmatrix} \tag{3.7}$$

If a and b are scalor and $g = ab$, then a can be recovered from g by dividing g by b, or b is recovered by dividing g by a. On the other hand, when a and b are matrices, there is no straightfoward division of a matrix by another matrix. The procedure of devision of matrices is a little more involved. Namely, to recever a from the product $g=ab$, g has to be postmultiplied by inverse of b, namely b^{-1}. To recover b from the product g, g must be premultiplied by inverse of a, namely a^{-1}.

Example

Assume

$$a = \begin{pmatrix} 2 & 1 \\ 1 & 2 \end{pmatrix}$$ (3.8)

$$b = \begin{pmatrix} 3 & 1 \\ 2 & 3 \end{pmatrix}$$

Then,

$$g = ab = \begin{pmatrix} 8 & 5 \\ 7 & 7 \end{pmatrix}$$ (3.9)

Using Octave/Matlab, inverses of a and b are respectively

$$a^{-1} = \begin{pmatrix} 0.66667 & -0.33333 \\ -0.33333 & 0.66667 \end{pmatrix}$$ (3.10)

$$b^{-1} = \begin{pmatrix} 0.42857 & -0.14286 \\ -0.28571 & 0.42857 \end{pmatrix}$$ (3.11)

Then, premultiplication of g by a^{-1} becomes

$$a^{-1}g = \begin{pmatrix} 3 & 1 \\ 2 & 3 \end{pmatrix}$$ (3.12)

which is b, and post multiplication of g by b^{-1} becomes

$$gb^{-1} = \begin{pmatrix} 2 & 1 \\ 1 & 2 \end{pmatrix}$$ (3.13)

which is a.

Finding solution of linear equations
A set of simultaneous linear equations

$$c_{1,1}x_1 + c_{1,2}x_2 + ... + c_{1,n}x_n = y_1$$

$$c_{2,1}x_1 + c_{2,2}x_2 + \ldots + c_{2,n}x_n = y_2$$

$$\ldots \tag{3.14}$$

$$c_{n,1}x_1 + c_{n,2}x_2 + \ldots + c_{n,n}x_n = y_n$$

may be written using matrix notations as

$$cx = y \tag{3.15}$$

where c is the coefficient matrix, and x and y are column vectors:

$$c = \begin{pmatrix} c_{1,1} & c_{1,2} & \cdots & c_{1,n} \\ c_{2,1} & c_{2,2} & \cdots & c_{2,n} \\ .. & .. & \cdots & .. \\ c_{n,1} & c_{n,2} & \cdots & c_{n,n} \end{pmatrix}, \quad x = \begin{pmatrix} x_1 \\ x_2 \\ \ldots \\ x_n \end{pmatrix}, \quad y = \begin{pmatrix} y_1 \\ y_2 \\ \ldots \\ y_n \end{pmatrix} \tag{3.16}$$

The solution of the linear equation is calculated in Octave/Matlab by

```
>>x=c\y
```

Example

$$\begin{pmatrix} 3 & 1 \\ -1 & 2 \end{pmatrix} \begin{pmatrix} x_1 \\ x_2 \end{pmatrix} = \begin{pmatrix} 0 \\ 1 \end{pmatrix}$$

```
>>c=[3 1; -1 2]; y=[0 1] '; x=c\y
x =
   -0.14286
    0.42857
```

Determinant

The determinant of a matrix a is computed by **det(a)** command. If the determinant is zero, there is no solution for the linear equation. For the 2-by-2 matrix in the earlier example, its determinant is computed as

```
>> c=[3 1; -1 2]; det(c)
ans =   7
```

Norm

The 2-norm of matrix c is computed by **norm(c)**:

```
>> c=[3 1; -1 2]; norm(c)
ans =   3.1926
```

Eigenvalue

Eigenvalues of matrix c is computed by **eig(c)** command:

```
>> c=[3 1; -1 2]; eig(c)
ans =
      2.5000 + 0.8660i
      2.5000 - 0.8660i
```

3.2 Polynomials and polynomial fitting

Polynomials in Octave/Matlab

In Octave/Matlab, a polynomial is expressed by an array of the power coefficients. For a polynomial in power series,

$$y = c_1 x^n + c_2 x^{n-1} + c_3 x^{n-2} + ... + c_{n+1} \qquad (3.17)$$

the array representing the polynomial is

$$c = [c_1, c_2, c_3, ..., c_{n+1}]$$

Polynomial of order *n* passign through *n*+1 data points

The polynomial of order n passing through $n+1$ data points is determined by **polyfit(x,y,length(x)-1)**.

Example
Data points are given:

x	y
0	1
1	-1
2	1

The polynomial passing through these three points is a polynomial of order 2:

xydata=[0 1; 1 -1; 2 1]
c=polyfit(xydata(:,1),xydata(:,2),2)
c =
    2.00000   -4.00000   1.00000

The foregoing result is interpreted as

$$y = 2x^2 - 4x + 1$$

Roots of a polynomial

Roots of a polynomial is computed by **roots(c)**, where c is the array of the power coefficients:

>>c=[2 -4 1]; roots(c)
ans =
    1.70711
    0.29289

Determining a polynomial using known roots

When all the roots of a polynomial are known as r, where r is an array of the roots, the polynomial can be determined by **poly(r)** except it becomes in the normalized form, which means that the leading coefficient is unity:

>>r=[1.70711,   0.29289]; c=poly(r)
c =
    1.00000   -2.00000   0.50000

Evaluation of a polynomial

For a polynomial with known coefficients c, the values of the polynomial can be evaluated by **polyval(c,x)** where x is a single value or an array of x values for which the polynomial is to be evaluated:

>>c=[2, -4, 1]; x=[1, 1.5, 2.3, 5]; polyval(c, x)
ans =
-1.00000   -0.50000   2.38000   31.00000

Differentiation of a polynomial

A polynomial represented by the coefficient array c is differentiated by **polyder(c)**. For example if $c=[2, -4, 1]$, >>polyder(c) yields
ans =
  4   -4
Does it make sense?

3.3 Numerical integration and differentiation

Trapezoidal rule

The trapezoidal rule is a numerical integration method to integrate a function. Suppose a function $f(x)$ is to be integrated in an interval of x, $a \le x \le b$. The interval is subdivided by grid points,

$$x_i = a + (i-1)h \tag{3.18}$$

where $h=(b-a)/n$, n is the number of subintervals, i is an integer, $i=1, 2, \dots n, n+1$. The functional values at the grid points are denoted by

$$f_i = f(x_i) \tag{3.19}$$

Then, the trapezoidal rule is

$$I = h(0.5f_1 + f_2 + f_3 + \dots + f_n + 0.5f_{n+1}) \tag{3.20}$$

An Octave/Matlab implementation of integration of $\exp(x)\cos(x)$ from $x=0$ to $x=1.4$ is illustrated next:

```
%Trapezoidal rule integration
disp('S is user-definition of the integrant: example ')
S= 'f=exp(-x).*cos(x); '
disp(' a, b, and n are user-input: following data are example ')
a = 0; b=1.4; n=30;
h=(b-a)/n; x=a+(0:n)*h;
eval(S);
disp('The result of integration ')
I=h*(sum(f) -0.5*(f(1)+f(length(f)) ))
```

Result:
$I = 0.60068$

Simpson's rule

The Simpson's rule is significantly more accurate than the trapezoidal rule. We use the same notations defined for the trapezoidal rule. The Simpson's rule needs at least two subintervals.

For the minimum value of n, namely $n=2$, the Simpson's rule is

$$I = (h/3)(f_1 + 4f_2 + f_3) \tag{3.21}$$

For a larger n, it is extended to

$$I = (h/3)(f_1 + 4f_2 + 2f_3 + 4f_4 + 2f_5 +.. \; 2f_{n-1} + 4f_n + f_{n+1}) \tag{3.22}$$

where n must be even.

```
%Simpson's rule integration
disp('S is user-definition of the integrant: example ')
S= 'f=exp(-x).*cos(x); '
disp(' a, b, and n are user-input: following data are example ')
a = 0; b=1.4; n=30;
h=(b-a)/n; x=a+(0:n)*h;
eval(S);
```

```
disp('The result of integration ')
I=sum(f)+3*sum(f(2:2:length(f)-1)) + sum(f(3:2:length(f)-2));
I=(h/3)*I
```

Result:
I = 0.60054

Numerical differentiation

Differentiation of a polynomial is derived by **polyder** command, but numerical computation of derivatives of all other functions needs a difference approximation.

Difference approximation uses the values of the function at discrete points near the point where the derivative is to be evaluated. Suppose the derivative is to be evaluated at $x=x_0$. For the first derivative, the backward difference, central difference and forward difference approximations are most frequently used.

Forward difference approximation:

$$f'(x_0) \approx (f(x_0+\Delta x) - f(x_0))/\Delta x \qquad (3.23)$$

Central difference approximation:

$$f'(x_0) \approx (f(x_0+\Delta x) - f(x_0-\Delta x))/2\Delta x \qquad (3.24)$$

Backward difference approximation:

$$f'(x_0) \approx (f(x_0) - f(x_0-\Delta x))/\Delta x \qquad (3.25)$$

where Δx is a small increment of x. For the second derivative, the central diffdrence approximation is most often used:

$$f''(x_0) \approx (f(x_0+\Delta x) - 2f(x_0) + f(x_0-\Delta x))/\Delta x^2 \qquad (3.26)$$

If the discrete points are separated by multiples of a unit interval, Δx, grid points may be expressed by $x_i = x_0 + i\Delta x$, where i is a negative or positive integer. The functional value at x_i, namely $f(x_i)$ is denoted by f_i. Then, the foregoing equations and some additional equations are written next:

$$f_i' \approx (f_{i+1} - f_i)/\Delta x$$

$$f_i' \approx (f_{i+1} - f_{i-1})/2\Delta x$$

$$f_i' \approx (f_i - f_{i-1})/\Delta x$$

$$f_i'' \approx (f_{i+1} - 2f_i + f_{i-1})/\Delta x^2$$

$$f_i'' \approx (-f_{i+2} + 16f_{i+1} - 30f_i + 16f_{i-1} - f_{i-2})/12\Delta x^2$$

$$f_i''' \approx (f_{i+2} - 2f_{i+1} + 2f_{i-1} - f_{i-2})/2\Delta x^3$$

There are two kinds of errors associated with any difference approximation. One is the truncation error, which decreases as Δx is decreased. Another is the roundoff error that becomes serious when Δx is too small. Because of these two kinds of errors, the total error of a difference approximation decreases with decrease of Δx down to a certain value of Δx beyond which the total error strats to increase as Δx is decreased further.

3.4 Finding roots of a nonlinear function

Roots of a polynomial may be found by **roots** command, but finding roots of nonlinear functions needs one of other methods, among which the Newton's iteration is most useful. However, Newton's iteration requires a decent initial guess for the root to find. To find an initial guess, graphic plotting of the function is recommended.

Suppose a function is denoted by $f(x)$. The roots of the function are the solutions of $f(x)=0$. Denoting an initial guess for the root as x_0, the equation is rewritten as

$$f(x_0 + \delta x)=0 \qquad (3.27)$$

where δx is a small unknown correction we have to find. Expanding the foregoing equation into a truncated Toylor series yields

$$f(x_0) + \delta x f'(x_0) \approx 0 \qquad (3.28)$$

Solution of the equtions for δx is

$$\delta x \approx -f(x_0)/f'(x_0) \qquad (3.29)$$

Because we solve the truncated Taylor expansion, $x_0 + \delta x$ thus calculated does not satisfy Eq.(3.27) exactly. Therefore, we set $x_1 = x_0 + \delta x$, and consider it as an improved estimate. We repeat the same procedure, namely

$$\delta x \approx -f(x_1)/f'(x_1) \qquad (3.30)$$

This procedure is iterated, namely $x_2 = x_1 + \delta x$ and

$$\delta x \approx -f(x_2)/f'(x_2)$$

so on, until the absolute value of δx becomes smaller than a prescribed criterion. In the foregoing equations, the first derivatives may be computed by a difference approximation,

$$f'(x) \approx (f(x + \Delta x) - f(x))/\Delta x \qquad (3.31)$$

where Δx is a small arbitrary value such as 0.001, but should not be too small because of the round off error of the difference approximation. The convergence rate of Newton's iteration is not sensitive to the errors of the difference approximation as long as Δx is reasonably small but not too small.

Example

We consider

$$f = \sin(x) + \exp(-0.4*x)$$

as a sample equation to solve. In the following program, the equation is defined as a string variable S. The initial guess x0 is set to 3.

```
%Newton's iteration
clear
disp(' ')
disp('User-definition of the equation to solve:   ')
S=' f= sin(x) + exp(-0.4*x) ; ' ; disp(S)
disp(' ')
x0=3;
disp(' Itr.No    Itr. Sol, x        Residue')
for n=1:10
x=x0;eval(S);f0=f;    x=x0+0.001; eval(S); f0pdx=f;
f_deriv=(f0pdx-f0)/0.001;
delx=-f0/f_deriv;
x=x0+delx;
fprintf('    %i    %.6e    %.2e \n', n,x,delx)
if abs(delx)<0.000001; break
end %if
x0=x;
end
disp('End of solution   ')
disp(' ')
```

Result

Itr.No	Itr. Sol, x	Residue
1	3.398296e+000	3.98e-001
2	3.401046e+000	2.75e-003
3	3.401047e+000	6.67e-007
End of solution		

The foregoing script may be adapted to other equations by changing S, and initial guess x0.

3.5 Boundary value problem of ordinary differential equation (ODE)

Consider a straight thin uniform metal rod of length L connected to a heat source of temperature T_L at the left end and another heat source of temperature T_R at the right end. The surface of the rod is exposed to air of temperature T_{air}. The x coordinate is the distance from the left end. We assume the temperature inside the rod is constant across any cross section of the rod.

The temperature of the rod $T(x)$ satisfies the differential equation:

$$-kAT'' + Ph(T(x) - T_{air}) = 0 \qquad (3.32)$$

where, A is the cross sectional area, k is the thermal conductivity, P is perimeter of the rod, h is the convection heat transfer coefficient, and T'' is the second derivative of $T(x)$. The temperature is subject to two boundary conditions:

$$T(0) = T_L$$
$$T(L) = T_R$$

Dividing Eq.(3.32) by kA yields

$$-T'' + eT = eT_{air} \qquad (3.33)$$

where $e = Ph/(kA)$.

To solve Eq.(3.33) by difference approximation, we consider a mesh defined by $x_i = i\Delta x$, $i = 0, 1, 2, \ldots n+1$, where $\Delta x = L/(n+1)$, and where $n+1$ is the number of mesh intervals between the left end to the right end of the rod. At the left end of the rod, i equals 0, and at the right end i equals $n+1$. We also use the notation $T_i = T(x_i)$. Therefore, $T_0 = T_L$ and $T_{n+1} = T_R$.

With the central difference approximation for T ", Eq.(3.33) is approximated by

$$- T_{i-1} + 2T_i - T_{i+1} + qT_i = qT_{air}, \quad i=1, 2, .. n \qquad (3.34)$$

with $q = e\,\Delta x^2$ and

$$T_0 = T_L, \quad T_{n+1} = T_R$$

Equation (3.34) is a set of n linear equations, so it can be written in a matrix form given by

$$mt = d \qquad (3.35)$$

where

$$m = \begin{pmatrix} 2+q & -1 & & \\ -1 & 2+q & -1 & \\ & & ... & \\ & & -1 & 2+q \end{pmatrix}, \quad t = \begin{pmatrix} T_1 \\ T_2 \\ .. \\ T_n \end{pmatrix}, \quad d = \begin{pmatrix} qT_{air} + T_L \\ qT_{air} \\ .. \\ qT_{air} + T_R \end{pmatrix}$$

Matrix m may be expressed in the form

$$m = \begin{pmatrix} b(1) & c(1) & & \\ a(2) & b(2) & c(2) & \\ & & ... & \\ & & a(n) & b(n) \end{pmatrix} \qquad (3.36)$$

which is named a *tridiagonal* matrix. The non-zero entries may also be expressed by array variables as follows

$a = [\,0, -1\ -1\ ...\ -1]$
$b = [2+q, 2+q, 2+q\ ...\ 2+q]$
$c = [-1, -1, -1, ...-1, 0]$
$d = [qT_{air} + T_L,\ qT_{air},\ ...\ qT_{air} + T_R]$

The foregoing equation in the matrix form may be solved by $t=m\backslash d$, but another method named the tridiagonal solution algorithm is better (computationally more efficient), which is explained next:

(a) Evaluate recurrently the following equations for i=2, 3, ... n in increasing order of i:

$r=a(i)/b(i-1)$
$b(i)= b(i)-r*c(i-1)$
$d(i)= d(i)-r*d(i-1)$

(b) Evaluate:

$d(n)= d(n)/b(n)$

(c) Recurrently evaluate the following equations in descending order of i:

$d(i)=(d(i)-c(i)*d(i+1))/b(i)$

The solution is in the array d, that is $T(i)=d(i)$. A user-defined **function** for the tridiagonal solution is as follows:

```
function f=tri_diag(a,b,c,d,n)
for i=2:n
  r=a(i)/b(i-1); b(i)=b(i)-r*c(i-1); d(i)=d(i)-r*d(i-1);
end %for
d(n)=d(n)/b(n);
for i=n-1:-1:1
  d(i)=(d(i)-c(i)*d(i+1))/b(i);
end %for;
f=d;
```

Example
Compute the temperature distribution in a steel rod for the two cases of heat transfer coefficients and plot the results in a graph:
Thermal conductivity: k=80W/mC (steel)
Heat transfer coefficient:
h=10w/m^{2}C (slow air flow)
h=300W/m^{2}C (high speed air flow)
Boundary conditions: TL=400C, TR=20C, Tair= - 10C
Metal rod of radius 0.01m, length L=0.3m

```
%ODE Boudary Value Problem
clear all, clf
k=80; h=300;TL=400; TR=20; Tair= -10; R=0.01;
A=pi*R^2; P=2*pi*R; L=0.2; n=100; dx=L/(n+1);
for kase=1:2
if kase==1; h=10;
elseif kase==2, h=300;
end %if
e=dx^2*P*h/k/A;
a(2:n)=-1;
b(1:n)=2+e;
c(1:n-1)=-1;
d(1:n)=e*Tair; d(1)=d(1)+TL; d(n)=d(n)+TR;
%Tridiagonal solution begines
for i=2:n
   r=a(i)/b(i-1); b(i)=b(i)-r*c(i-1);
   d(i)=d(i)-r*d(i-1);
end %for
d(n)=d(n)/b(n);
for i=n-1:-1:1
   d(i)=(d(i)-c(i)*d(i+1))/b(i);
end %for;
% End of tridiagonal solution
nhalf=ceil(n/4)
T=d;x(1:n)=(1:n)*dx;
plot(x,T); hold on
if kase==1, text(x(nhalf),T(nhalf)+20, ...
                'Low air flow', 'fontsize', 14 );
elseif kase==2, text(x(nhalf),T(nhalf)+20, ...
                'High air flow', 'fontsize', 14   );
end %if
end %for %case
xlabel('X meter', 'fontsize',16);
ylabel('T degrees C', 'fontsize',16)
```

Result

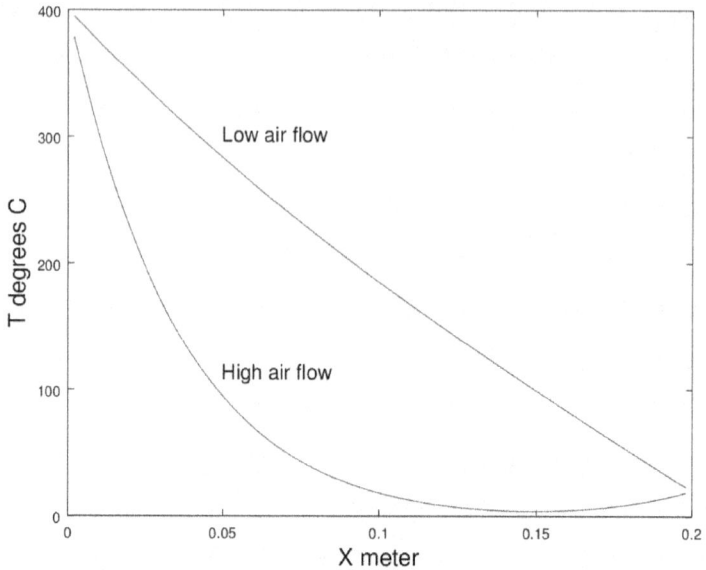

Figure 3.1   Temperature distributions in the rod

3.6 Initial value problem of ordinary differential equation (ODE)

1st order ODE

Consider a simple example of ordinary differential equation given by

$$y'(t) = f(y(t), t) , \quad \text{with } y(0) = y_0 \tag{3.37}$$

where t is time, c is a constant and y_0 is the known value of y at the initial time $t=0$, and y' is the first derivative of y with respect to t. Time-dependent ordinary differential equations are all initial value problems.

When using a numerical method based on a difference approximation we consider a grid on time axis, $t_i = i\Delta t$, where i is an integer, and the initial time is $t_0 = 0$.

We now consider a time interval, $t_i \leq t \leq t_{i+1}$, assuming the value of y_i has been calculated, or is known. By integrating Eq.(3.25) in this interval, we get

$$y_{i+1} = y_i + \int_0^{\Delta t} y'(t)dt = y_i + \int_0^{\Delta t} f(y,t)dt \qquad (3.38)$$

However, we realize that integrant $y'(t)$ or y inside the integral is unknown. Depending on how to approximate the integrant $y'(t)$, several different numerical methods can be derived.

We condider here only the 2nd order Runge-Kutta method, which is derived by using the trapezoidal rule integration method:

$$y_{i+1} \approx y_i + \Delta t(f_i + f_{i+1})/2 \qquad (3.39)$$

where f_{i+1} is approximated by

$$f_{i+1} \approx f(y_i + \Delta t f_i, t_{i+1}) \qquad (3.40)$$

By defining

$$k_1 = \Delta t\, f(y_i, t_i)$$
$$k_2 = \Delta t\, f(y_i + \Delta t f_i, t_{i+1})$$

the 2nd order Runge-Kutta method is written as

$$y_{i+1} = y_i + (k_1 + k_2)/2 \qquad (3.41)$$

2nd order ODE

The second order Runge-Kutta method can be applied to a higher-order ODEs. Here we explain for the case of a second-order ODE:

$$u''(t) + au'(t) + bu(t) = q(t) \qquad (3.42)$$

with initial conditions,

$$u(0) = u_0, \quad u'(0) = u_0'$$

Equation (3.42) may be reduced to first order differential equations by defining a new function $v(t)$ by

$$v(t) = u'(t)$$

Equation (3.42) becomes a set of coupled first order differential equations:

$$
\begin{aligned}
u' &= v, & u(0) &= u_0 \\
v' &= -av - bu + q, & v(0) &= u_0'
\end{aligned}
\qquad (3.43)
$$

The foregoing equation may be written as a single equation using the matrix notations as

$$y' = f \qquad (3.44)$$

where

$$y = \begin{pmatrix} u \\ v \end{pmatrix}, \qquad f = \begin{pmatrix} v \\ -av - bu + q \end{pmatrix}$$

The 2nd order Runge-Kutta method applied to this matrix form is written as

$$
\begin{aligned}
k_1 &= \Delta t\, f(y_i, t_i) \\
k_2 &= \Delta t\, f(y_i + \Delta t f_i, t_{i+1}) \\
y_{i+1} &= y_i + (k_1 + k_2)/2
\end{aligned}
\qquad (3.45)
$$

where k_1 and k_2 are vertical vectors.

Example

A rectangular object of mass M=5kg is fixed to the lower end of a massless spring-damper system as illustrated in Figure 3.2. The upper end of the spring is fixed to a structure at rest. The box receives a force of R= -Bu' from the damper when moving at velocity u'(t), where B is a damping constant. The equation of motion is

$$Mu'' + Bu' + ku = 0, \quad u(0)=1, u'(0)=0$$

where u is displacement from the static position, k is the spring constant equal to 100N/m, and $B=5$Ns/m. Calculate $u(t)$ for $0<t\leq5$s using the 2nd order Runge Kutta method.

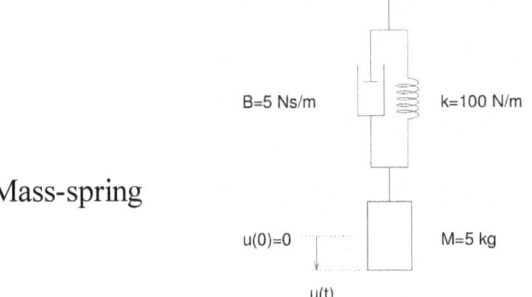

Figure 3.2 Mass-spring system

Solution
The equation of motion may be written as

$$y' = my$$

where

$$y = \begin{pmatrix} u \\ v \end{pmatrix}, \quad m = \begin{pmatrix} 0 & 1 \\ -b & -a \end{pmatrix}$$

with $b=B/M=5/5=1$ Ns/mkg, $a=k/M=100/5=20$ N/kg. Here, v is the velocity which equals u'.

The results of the computation are shown in the figure below. The program is listed below.

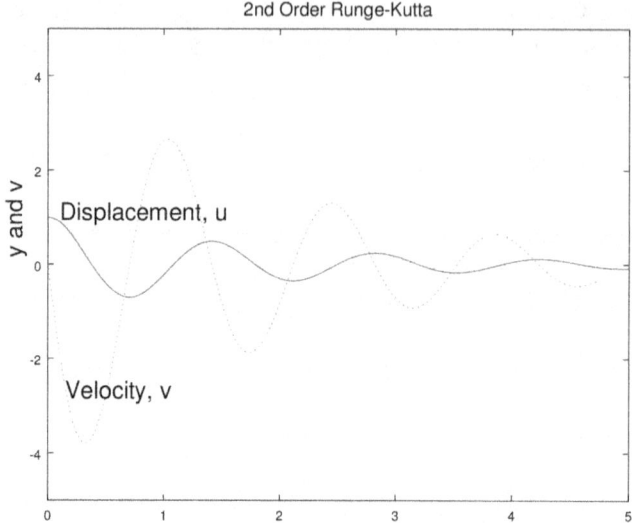

Figure 3.3 Result of the 2nd order Runge-Kutta method

Figure 3.3 Result of the 2nd order Runge-Kutta method

```
%ODE Initial Value Problem
clear,clf
M=5; k = 100; B=5; a = B/M;
b = k/M; n=0; t=0; h = 0.025;
y(:,1) = [1;0]; t(1)=0; % initial condition
m=[0,1; -b,-a]; tmax=5;
while t<=tmax
  n=n+1;
   k1 = h*m*y(:,n);
   k2 = h*m*(y(:,n)+k1);
   y(:,n+1) = y(:,n) + (k1 + k2)/2;
   t(n+1) = n*h;
end
plot(t,y(1,:), '-', t,y(2,:),':');
h1=text(0.6513,   -17.8152,'t (s)');
set(h1,'FontSize',[18])
h2=text(-0.3247,   0, ' y and v');
set(h2,'FontSize',[18],'Rotation',[90])
text(t(5),y(1,5)+0.3,'Displacement, u','FontSize',[18])
text(t(7),y(2,7),'Velocity, v','FontSize',[18])
axis([0,tmax,-5,5])
title('2nd Order Runge-Kutta','fontsize',14)
```

Although we did not explain the 4th order Runge-Kutta method, the 2nd order Runge-Kutta method in the program can be switched by rewriting only the while/end loop in the program to:

```
while t<=tmax
  n=n+1;
    k1 = h*m*y(:,n);
    k2 = h*m*(y(:,n)+k1/2);
    k3 = h*m*(y(:,n)+k2/2);
    k4 = h*m*(y(:,n)+k3);
    y(:,n+1) = y(:,n) + (k1 + 2*k2 + 2*k3 + k4)/6;
    t(n+1) = n*h;
  end
```

The results of the 4th order Runge-Kutta method is shown in the following figure.

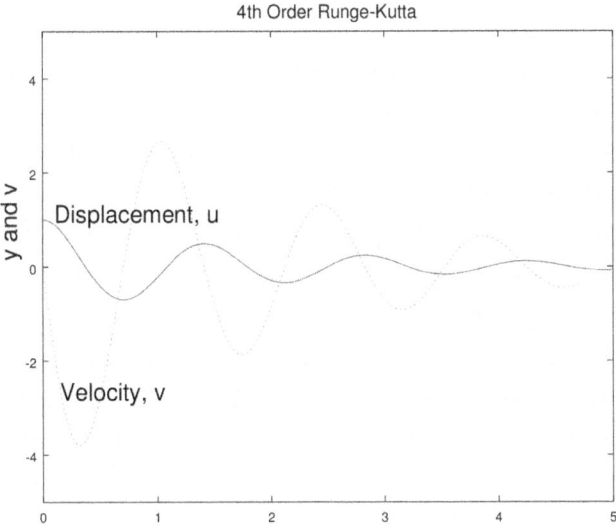

Figure 3.3   Result of the 4th order Runge-Kutta method

3.7 Parabolic partial differential equation (PDE)

Consider a heated object, such as a mug of very hot soup or a potato that is taken out of microwave oven at $t=0$. We have experience of waiting 5 to 10 minutes until the the object becomes cooler. Also, in case of soup in a mug, if viscosity of soup is increased by some means (for example by mixing cooked corn starch) the speed of cooling is significantly slowed down. Investigating how the temperature inside the object cools down is a bit of science.

As a mathematical model, we consider an infinitely long cylinder of a solid material and use one-dimentional cylindrical coordinate, r (we will consider a liquid material later). We assume the outer surface is exposed to air of temperature T_{air}, where convection heat transfer occurs. The temperature $T(r, t)$ is a function of r and time t.

The whole problem is written mathematically as follows. The temperature inside the cylinder of a solid material satisfies the parabolic partial differential equation given by

$$\rho c \frac{\partial T(r,t)}{\partial t} - \frac{1}{r}\frac{\partial}{\partial r} kr \frac{\partial T(r,t)}{\partial r} = 0 \qquad (3.46)$$

where ρ is the density of the material, c the specific heat, k the thermal conductivity. We assume that k is constant, so k can be moved out of the differential operator. The foregoing equation needs two boundary conditions,

$r=0$: $\partial T / \partial r = 0$
$r=R$: $k\partial T / \partial r = h(T_{air} - T(R))$

and one initial condition,

$$T(r,0) = T_{initial}(r)$$

where $T_{initial}$ is the known initial temparature distribution at $t=0$.

To discretize the equation, we consider mesh points in time as well as in space. The time mesh points are $t_n=n\Delta t$, the space mesh points are $r_i=i\Delta r$, $i=0, 1, 2, 3, \ldots$ imax, with $\Delta r=R/imax$, and where R is the radius of the cylinder. On the time axis, we use the backward difference approximation, namely

$$\frac{\partial T(r,t)}{\partial t} \approx \frac{T_i^n - T_i^{n-1}}{\Delta t} \tag{3.47}$$

and use the central difference approximation on the space mesh:

$$\frac{k}{r}\frac{\partial}{\partial r} r \frac{\partial T(r,t)}{\partial r} = \frac{k}{r_i}\left(r_{i+1/2}T_{i+1/2}^{n}{}' - r_{i-1/2}T_{i-1/2}^{n}{}'\right)/\Delta r \tag{3.48}$$

with

$$T_{i+1/2}^{n}{}' = (T_{i+1}^n - T_i^n)/\Delta r$$
$$T_{i-1/2}^{n}{}' = (T_i^n - T_{i-1}^n)/\Delta r$$

Putting all the terms together, we get

$$-\frac{kr_{i-1/2}}{\Delta r}T_{i-1}^n + (\frac{\Delta r \rho c r_i}{\Delta t} + \frac{2kr_i}{\Delta r})T_i^n - \frac{kr_{i+1/2}}{\Delta r}T_{i+1}^n$$
$$= \frac{\Delta r \rho c r_i}{\Delta t}T_i^{n-1} \tag{3.49}$$

$$\text{for } i=2, 3, \ldots i_{max}-1$$

where $2r_i = r_{i+1/2} + r_{i-1/2}$ is used. For $i=0$, which is the center of sphere, $r_0=0$, we can use

$$T_0 = T_1$$

as the boundary condition. Therefore, Eq.(3.49) for $i=1$ becomes

117

$$(\frac{\rho c r_1}{\Delta t} + \frac{k r_{1+1/2}}{\Delta r})T_1^n - \frac{k r_{1+1/2}}{\Delta r}T_2^n = \frac{\rho c r_1}{\Delta t}T_1^{n-1} \tag{3.50}$$

For i=imax, we rewrite Eq.(3.48) to

$$\frac{k}{r}\frac{\partial}{\partial r}r\frac{\partial T(r,t)}{\partial r} = \frac{k}{r_{i\max}}\left(r_{i\max}T_{i\max}{}' - r_{i\max-1/2}T_{i\max-1/2}{}'\right)\frac{2}{\Delta r} \tag{3.51}$$

We also rewrite the boundary condition for the outer surface to

$$T_{i\max}{}' = (h/k)(T_{air} - T_{i\max}) \tag{3.52}$$

Substituting Eq.(3.52) to Eq.(3.51) yields

$$\frac{k}{r}\frac{\partial}{\partial r}r\frac{\partial T(r,t)}{\partial r} =$$

$$\left(h(T_{air} - T_{i\max}) - \frac{k}{r_{i\max}}r_{i\max-1/2}\frac{T_{i\max} - T_{i\max-1}}{\Delta r}\right)\frac{2}{\Delta r} \tag{3.53}$$

Combining Eq.(3.46), Eq.(3.47) and Eq.(3.53) together yields

$$-\frac{k}{\Delta r}r_{i\max-1/2}T_{i\max-1}^n + (\frac{\Delta r \rho c r_{i\max}}{2\Delta t} + hr_{i\max} + \frac{k}{\Delta r}r_{i\max-1/2})T_{i\max}^n$$

$$= hr_{i\max}T_{air} + \frac{\Delta r \rho c r_{i\max}}{2\Delta t}T_{i\max}^{n-1} \tag{3.54}$$

where the equation has been multiplied by $r_{i\max}\Delta r/2$. Given the values of T_i^{n-1}, Eqs. (3.50), (3.49) and (3.54) become a set of tridiagonal equations in each time step, which is solved by the tridiagonal solution method explained in Section 3.5.

Numerical experiments
Case 1

118

We assume a cylinder of radius 0.05m filled with water. We also assume that the cylindrical vessel is infinitely thin, ignore its effect to heat transfer for simplicity and use the following cons-tants:

Water density(ro)=1000kg/m³
Specific heat(cp) =4179 J/kgC
Thermal conductivity(k)=0.6 W/mC
Heat transfer coefficient (h at cylindrical surface)=100W/m²C
Air temperature outside the cylindr(T_{air})=20C
Initial temperature of water(T_{ini})=100C
Δt=60 sec

We make one assumption: actual water circulates in a vessel causing thermal convection, which cannot be simulated by the present model. Therefore, we assume that water does not circulate by natural convection. This is the case if the water is well mixed with cooked corn starch.

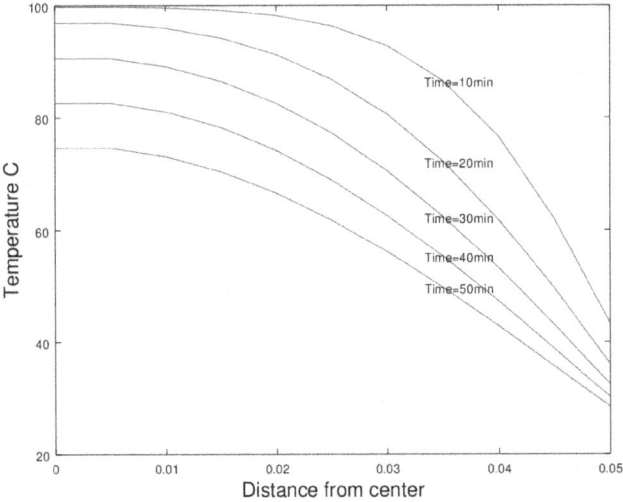

Figure 3.4 Temperature changes with no thermal convection

Then the water temperature change is simulated by the program that solves the equation developed. The program is listed later. The results are shown in Figure 3.4. The speed of cooling down seems to be slower than what we know with cooling of

119

coffee in a mug. There are three reasons. First, the diameter of the cylinder is larger than usual coffee mugs. Second, an infinitely long cylinder is the model while actual coffee mug has a finite length. Third, natual convection in the cylinder is totally ignored.

Case 2

We wish to simulate the case with water circulation in the cylinder by natural convection or even forced convection. Such mixing effect may be qualitatively emulated by assuming the thermal conductivity is artificially increased. So we increase k by a factor of 10. The result is shown in Figure 3.5.

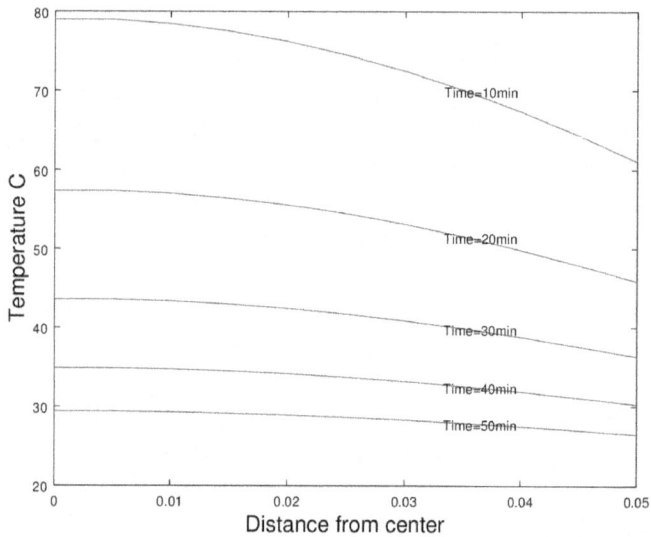

Figure 3.5 Temperature change when thermal conductivity is increased by a factor of 10

```
%Parabolic PDE
clear, clf
k=0.6*10;        %W/mC
ro=1000;       % kg/m^3density
cp=4.179*1000;     %J/kgC Specific heat of water
h=100;          %W/m^2C   Moderaate speed of air flow
R=0.05;         %diameter of cylinder, m
imax=10;         %maximum mesh index
Tair=20;       % degrees C
```

```
Tini=100;    %   Initial internal temperature degrees C
dt=60;          %s time interval for calculatios
dr=R/imax;
rr=dr*(1:imax);
% Geometrical note I=0 is the center; i=1 is next point not center
for i=1:imax-1
aa(i)=-k*(rr(i)-dr/2)/dr;
bb(i)=dr*ro*cp*rr(i)/dt+2*k*rr(i)/dr;
cc(i)=-k*(rr(i)+dr/2)/dr;
dcoef(i)=dr*ro*cp*rr(i)/dt;
endfor
bb(1)=bb(1)+aa(1);
aa(imax)=-k*(rr(imax)-dr/2)/dr;
bb(imax)=( 0.5*dr*ro*cp*rr(imax)/dt + h*rr(imax)+k*(rr(imax)-
dr/2)/dr);
cc(imax)=0; %dummy

time=0;
%Initial temperature condition
TB(1:imax)=Tini;
%for case=1:2
for timestep=1:50
time=time+dt;
a=aa;b=bb;c=cc;
for i=1:imax-1
d(i)=dcoef(i)*TB(i);
end
d(imax)=h*rr(imax)*Tair+0.5*dr*ro*cp*rr(imax)/dt*TB(imax);
%Tridiagonal solution begins
n=imax;
for i=2:n
   r=a(i)/b(i-1); b(i)=b(i)-r*c(i-1);
   d(i)=d(i)-r*d(i-1);
endfor
d(n)=d(n)/b(n);
for i=n-1:-1:1
   d(i)=(d(i)-c(i)*d(i+1))/b(i);
endfor;
%end of tridiagonal solution
T=d;
if mod(timestep,10)==0
plot([0,rr,R],[T(1),T,Tair]); hold on
s=['Time=',num2str(time/60),'min']
rtx=R*2/3; Ttx=T(floor(rtx/dr)+1);
text(rtx,Ttx,s );
```

```
end %if
TB=T;
end% timestep
ylabel('Temperature C', 'fontsize',16)
xlabel('Distance from center','fontsize', 16)
```

Comparing Figure 3.5 to Figure 3.4, we observe that the cooling speed is significantly increased by the increase of the thermal conductivity, which emulates in our model the effect of convection of water in the cylinder.

Exercise problems for Chapter 3

[1] Two matrices and one vector are defined by

$$a = \begin{pmatrix} 3 & 1 \\ 0 & 2 \end{pmatrix}, \quad b = \begin{pmatrix} 5 & 1 \\ -2 & 1 \end{pmatrix}, \quad v = \begin{pmatrix} 2 \\ 7 \end{pmatrix}$$

Calculate ab, av, and $b^{-1}v$.

[2] Define two matrices by $a=[1\ 2;3\ 4]$ and $b=[3\ 1;2\ 4]$. Calculate $g=a*b$, $inv(a)*g$, $g*inv(b)$, $g*inv(a)$, and $inv(b)*g$.

[3] Solve the following linear equations. Calculate the determinant of the coefficient matrix for each also:

    (a)
$$2x + y + z = 4$$
$$x - y - 5z = -5$$
$$-2x + 2y + z = 1$$

    (b)
$$-a + b + c - d = 3$$
$$a + 2b - 3c + 3d = -8$$
$$-2a + 2b + c + 2d = -1$$
$$a + 5b - c - 3d = 7$$

(c)

$$2a + b + 4c + d = 4$$
$$a + 2b - 3c + 3d = -1$$
$$-2a - 3b - c + 2d = 7$$
$$-a - b - 4c + 5d = 8$$

[4] A polynomial passes through the five points, $x=[0\ 1\ 2\ 3\ 4]$, $y=[0\ 1\ 0\ 0\ 0]$. Determine the coefficients of the polynomial in power series. Make sure the polynomial determined satisfies the points by **polyval**.

[5] Plot the polynomial determined for [4] together with its derivative in $-1 \leq x \leq 5$. Do not forget to add xlabel, ylabel, and legends.

[6] Evaluate the following integers by Simpson's rule with $n=2, 4, 8, 16$ subintervals:

(a) $\int_0^1 x \exp(2x)dx$

(b) $\int_0^1 x^{-x}dx$

(c) $\int_1^2 \frac{\log_e(1+x)}{x}dx$

[7] Find a root of the following equation near $x=3$ by Newton's iteration:

$$\sin(x) \exp(-0.4x) = 0$$

[8] Steady-state heat conduction equation on the cylindrical coordinate is

$$(k/r)(rT'(r))' = 0$$

where r is the radius coordinate and symbol ' is the different-tiation with respect to r, and k is the thermal conductivity. Suppose a cylindrical shell with inner radius 0.1m and outer radius 0.15m. Inside the inner surface is maintained at a constant temperature of 100C and the outer surface is exposed to air of 20C. The material of the cylindrical shell is steel or aluminum. The thermal conductivity of steel and aluminum are: $k=55$ W/mC for steel and $k=250$ W/mC for alminum. The heat transfer coefficient at the outer wall is 200 %W/m²C

The boundary conditions are:

$$T(0.1)=100$$
$$h(T(0.15)- T_{air})= - kdT/dr \quad \text{at } r=0.15$$

Determine the temperature distribution for each case of (1) steel, and (2) aluminum.

[9] We now revisit the problem of parabolic partial differential equation for the transient temperature distribution in a cylindrical shell described in Section 3.6. We assume the cylinder has been refrigerated at 0C, and suddenly placed in boiling water of 100C at $t=0$. Run the program developed in Section 3.7 after appropriate change of parameter(s), and plot how the temperature inside the cylinder change in time. Assume the heat transfer coefficient at the outer surface is $h=1000$W/m²C.

[10] Although we used a cylindrical model in Section 3.7, we may switch to a spherical model. The only change necessary is that the heat conduction equation for a spherical coordinate is

$$\rho c \frac{\partial T(r,t)}{\partial t} - \frac{1}{r^2} \frac{\partial}{\partial r} kr^2 \frac{\partial T(r,t)}{\partial r} = 0$$

Rewrite the program based on the foregoing equation. Solve the same problem using the spherical model, and investigate the difference in both solutions.

Chapter 4
Application Programs for Chores and Games

Think for a moment where in industry powerful computers are extensively utilized. They are not for solving mathematical equations but in many other non-mathematical areas. Some examples are: geogolical, medical and stellite image processing; business data processing; economic modeling; banking systems; insurance companies; logistic decision making; military and airforce; weather forecast; digital security; information service; entertainment industry; artificial intelligence; hospitls among many others. As one example, Octave was used on a massive parallel computer at Pittsburgh supercomputing center to find vulnerabilities related to guessing social security numbers.

It is said that those who have knowledge of programming for these areas will never struggle to find job opprtunities in the next two decades at least. If the reader is interested in developing personal capability in programming for any of these areas, the first step would be to start writing programs to do chores and games.

With these prospects in view, this chapter shows short applications of Octave/Matlab to do some chores and games.

4.1 Bubble sort

Bubble sort is a simple iterative procedure by which numbers in an array are reordered in ascending or descending order. In each iteration, the work starts from the top of the array and ends at the bottom. In the ascending reordering, each number is compared to the next in the array and if greater than the next number, they are swapped. In case of descending sort, swapping takes place if the number is less than the next one. The iteration is continued until no more swapping is necessary.

126

In the following script, an array of numbers, V, is sorted by the bubble sort algorithm for ascending order.

```
clc
clear all
disp('This program illustrates the bubble sort that ')
disp('reorders an array of numbers in ascending order.')
disp ('INPUTS')
disp('Input the array of numbers')
V=[18   7   6   15   4   13 ];
disp(V)
%% SOLUTION
% Number of entries, n
n=length(V);
k=0;
while 1
    k=k+1;
    %   Each number is compared with the next and swapped
    %   if the next is greater.
    noex=0; %Counter of swapping is initialized
    for i=1:1:n-1
    if V(i)>V(i+1);
        temp=V(i);
        V(i)=V(i+1);
        V(i+1)=temp;
        noex=noex+1;
    end %if
    end %for
    % Iteration ends if there is no swapping
    if noex==0, break
    end %if
end
disp('Number of iterations'), k
%% OUTPUT
disp(' ')
disp ('OUTPUT')
disp ('The ascending array is')
disp(V)
```

The output of the foregoing script is:

This program illustrates the bubble sort that
reorders an array of numbers in ascending order.

INPUT
Input array of numbers
    18   7   6   15   4   13
Number of iterations
k = 5

OUTPUT
The ascending array is
    4   6   7   13   15   18

The same chore may be done by the **sort** command as follows:

```
>>V=[18  7  6  15  4  13]';
>> sort(V)
ans =
      4
      6
      7
     13
     15
     18
```

By using the **sort** command as follows, both the result of the sorting and the initial order are printed out:

```
>>[a, i]=sort(V)
[a,i]=sort(V)
a =
      4
      6
      7
     13
     15
     18

i =
      5
      3
      2
      6
      4
      1
```

Here a is the result of sorting and i is the initial order. For example, i=5 at the top means that the top number 4 of a was the 5th entry before sorting.

The array i is useful when rows of a table are sorted in increasing/decreasing order of numbers in a selected column. Look at the following table:

18	33.1	1	0.12
7	12.25	3	0.34
6	105.89	2	1.55
15	41.00	0	3.21
4	6.96	5	0.77
15	3.44	3	0.81

Suppose we wish to reorder the rows in increasing order of the numbers in the first column. The remainder of the work is done in the following script:

```
T=[
    18   33.1      1   0.12;
     7   12.25     3   0.34;
     6   105.89    2   1.55;
    15   41.00     0   3.21;
     4   6.96      5   0.77;
    15   3.44      3   0.81
];
[b,i]=sort(T(:,1))
for k=1:length(i)
S(k,:)=T(i(k),:);
end
S
```

Result:
S =

4.00000	6.96000	5.00000	0.77000
6.00000	105.89000	2.00000	1.55000
7.00000	12.25000	3.00000	0.34000
15.00000	41.00000	0.00000	3.21000
15.00000	3.44000	3.00000	0.81000
18.00000	33.10000	1.00000	0.12000

The sorting may also be done in increasing order of the 2nd column by changing **[b,i]=sort(T(:,1))** to **[b,i]=sort(T(:,2))**:

```
%T=[ ...same...]
[b,i]=sort(T(:,2))
for k=1:length(i)
S(k,:)=T(i(k),:);
end
S
```

Result:
S=

15.00000	3.44000	3.00000	0.81000
4.00000	6.96000	5.00000	0.77000
7.00000	12.25000	3.00000	0.34000
18.00000	33.10000	1.00000	0.12000
15.00000	41.00000	0.00000	3.21000
6.00000	105.89000	2.00000	1.55000

To sort in descending order, change **[b,i]=sort(T(:,2))** to **[b,i]=sort(T(:,2), "descend")**.

6.00000	105.89000	2.00000	1.55000
15.00000	41.00000	0.00000	3.21000
18.00000	33.10000	1.00000	0.12000
7.00000	12.25000	3.00000	0.34000
4.00000	6.96000	5.00000	0.77000
15.00000	3.44000	3.00000	0.81000

Sorting may be done in alphabetic order by converting alphabetic characters to numbers. Consider the following table to be reordered in alphabetic order of the first column:

apple	18	33.1	1	0.12
orange	7	12.25	3	0.34
grape	6	105.89	2	1.55
pear	15	41.00	0	3.21
cherry	4	6.96	5	0.77
banana	15	3.44	3	0.81

The only way for Octave/Matlab to read the table, which a mixture of numbers and letters, is if all the contents are in string and all the rows are in the same length. So, in writing a script, we first convert the table to string. Then, letters are converted to numbers. The sorting work is done by the following program named "alphabetic_sort.m.

Program:

```
%alphabetic_sort.m
clear all
disp('Original Table')
T=[
'apple     18  33.1      1   0.12';
'orange     7  12.25     3   0.34';
'grape      6  105.89    2   1.55';
'pear      15  41.00     0   3.21';
'cherry     4  6.96      5   0.77';
'apple b   16  33.1      1   0.12';
'banana    15  3.44      3   0.81'
]
alpb=' abcdefghijklmnopqrstuvwxyz';   %27 characters
%Converting alphabet in column 1 to 8 to numbers
[high,wide]=size(T);
tr(1:high,1:9)=0;
tr(1:high,9)=[1:high]';
for i=1:high
   for j=1:8
      for k=1:27
         if T(i,j)==alpb(k), tr(i,j)=k; end %if
      end
   end
end

tr(1:high,9)=[1:high]'; % column 9 is added
%Sorting in order of column 8 to 1 in descending order
for L=8:-1:1
   [b,i]=sort(tr(:,L));
   for k=1:length(i)
      tr_tempo(k,:)=tr(i(k),:);
   end
   tr=tr_tempo;
end
disp('  ')
```

```
disp('Table after sorting in alphabetic order')
disp('=============================================')
for i=1:high
disp(T(tr(i,9),:))
end
disp('=============================================')
disp('END')
save -ascii file_name.txt T
```

Output:

```
>> alphabetic_sort
Original Table
T =
apple     18   33.1      1   0.12
orange     7   12.25     3   0.34
grape      6   105.89    2   1.55
pear      15   41.00     0   3.21
cherry     4   6.96      5   0.77
apple     16   33.1      1   0.12
banana    15   3.44      3   0.81

Table after sorting in alphabetic order
===========================================
apple     18   33.1      1   0.12
apple     16   33.1      1   0.12
banana    15   3.44      3   0.81
cherry     4   6.96      5   0.77
grape      6   105.89    2   1.55
orange     7   12.25     3   0.34
pear      15   41.00     0   3.21
===========================================
```

4.2 Calender maker

Making the calendar of any given year of this century takes some effort because what day in the week the first day of each month is needs to be found for any given year. Also, formatting monthly calendars is a quite challenging task with the rather primitive formatting capability of Octave/Matlab.

The following program performs this task nonetheless. The only one input is the year between 2016 and 2099 to be given as input through keyboard. The calendar created is saved in the file named "calender_xxxx.txt" where xxxx is the year number.

```
clear all, clf
day_of_week(1:100)=0;
for yr=2001:2099;
NOD='NOD_yr(yr-2000)=365;if mod(yr-2016,4)==0, NOD_yr(yr-2000)=366; end %if';
eval(NOD);
year(yr-2000)=yr;
end %for
%Seed information
% Jan 1 of 2016 is FRIDAY
day_of_week(16)=6;
for yr=17:99
day_of_week(yr)=mod(NOD_yr(yr-1)+day_of_week(yr-1),7);
if day_of_week(yr)==0, day_of_week(yr)=7;end %if
end %for

ind=0;
while ind+1
yr=input('Type Year Number between 2016 and 2099:    ')
if yr>2015 & yr<2100, ind=1; end %if
if ind==1, break; end %if
disp('You input is invalid. Please repeat')
end %while
yr=yr-2000;
disp('*---------*---------*---------*---------*---------*---------*')
%[year', NOD_yr', day_of_week']
if NOD_yr(yr)==365,
   NOD_of_mo=[31,28,31,30,31,30,31,31,30,31,30,31];
end %if
if NOD_yr(yr)==366,
   NOD_of_mo=[31,29,31,30,31,30,31,31,30,31,30,31];
end %if
%Number of day before the month
NOD_b4_mo(1)=0;
for i=2:12
NOD_b4_mo(i)=NOD_b4_mo(i-1)+NOD_of_mo(i-1);
end %for
calom(1:12,1:42)=0;
for mo=1:12;
  day_of_wk_of_1stday=mod(mod(NOD_b4_mo(mo),7)+day_of_week(yr), 7);
  if day_of_wk_of_1stday==0, day_of_wk_of_1stday=7;end %if
box(1:6,1:7)=0;
  k=0;
  n=0;
  for i=1:6
    for j=1:7
      k=k+1;
      n=n+1;
      if k<day_of_wk_of_1stday, n=0;
      end
      m=n;
      if n>NOD_of_mo(mo), m=0;
      end %if
      box(i,j)=m;
      calom(mo,k)=m;
```

```
            if i>4 & n<10, box(i,j)=0; calom(mo,k)=0;end %if

%       [i,j,k,n]
      end
    end
    %fprintf('MONTH %i\n', mo)
    %printing the calendar by box
    if box(6,1)==0, box(1:5,:);
    elseif box(6,1)~=0, box(:,:);
    end
  end %for
  calom;
% Final formatting

string(1:60,1:70)=' ';
s9='  Su Mo Tu We Th Fr Sa      Su Mo Tu We Th Fr Sa';
string(3,1:length(s9))=s9;
yrp=['    Year ',num2str(2000+yr)];
string(1,1:length(yrp))=yrp;
for mo=1:12
for k=1:42;%
s=calom(mo,k);
j=mod(k-1,7)+1;

i=ceil(k/7);
sH=floor(s/10);
sL=s-sH*10;
%disp('[k,j,i,s,sH,sL]')
%[k,j,i,s,sH,sL]
if sH==0 str(1)=' ';
    else str(1)=num2str(sH);
end %if

if sL==0 & sH==0, str(2)=' ';
else str(2)=num2str(sL);
end %if
me=mo;    if(mo>6) me=mo-6;    end %if
ie=i+(mod(me-1,7)+1)*8 -5;
je=j*4 ;
if mo>6 je=je+31; end %if
%[mo,ie,je]
string(ie,je)=str(1);
string(ie,je+1)=str(2);
%
    if k==1; monstr=['     Month ',num2str(mo)];
    bi=0; if mo>6, bi=31; end %if
      string(ie-2,1+bi:length(monstr)+bi)=monstr;
      string(ie-1,1:length(s9))=s9;
    end %if

end %for
end %for

for i=1:50
disp(string(i,:))%, fprintf('\n')
end %for
disp('*---------*---------*---------*---------*---------*---------*')
```

Type Year Number between 2016 and 2099:   2017
yr =   2017

Result

```
*——————*——————*——————*——————*——————*——————*

Year 2017
Month 1                          Month 7
Su  Mo  Tu  We  Th  Fr  Sa       Su  Mo  Tu  We  Th  Fr  Sa
 1   2   3   4   5   6   7                                1
 8   9  10  11  12  13  14        2   3   4   5   6   7   8
15  16  17  18  19  20  21        9  10  11  12  13  14  15
22  23  24  25  26  27  28       16  17  18  19  20  21  22
29  30  31                       23  24  25  26  27  28  29
                                 30  31
Month 2                          Month 8
Su  Mo  Tu  We  Th  Fr  Sa       Su  Mo  Tu  We  Th  Fr  Sa
             1   2   3   4                    1   2   3   4   5
 5   6   7   8   9  10  11        6   7   8   9  10  11  12
12  13  14  15  16  17  18       13  14  15  16  17  18  19
19  20  21  22  23  24  25       20  21  22  23  24  25  26
26  27  28                       27  28  29  30  31

Month 3                          Month 9
Su  Mo  Tu  We  Th  Fr  Sa       Su  Mo  Tu  We  Th  Fr  Sa
             1   2   3   4                            1   2
 5   6   7   8   9  10  11        3   4   5   6   7   8   9
12  13  14  15  16  17  18       10  11  12  13  14  15  16
19  20  21  22  23  24  25       17  18  19  20  21  22  23
26  27  28  29  30  31           24  25  26  27  28  29  30

Month 4                          Month 10
Su  Mo  Tu  We  Th  Fr  Sa       Su  Mo  Tu  We  Th  Fr  Sa
                         1        1   2   3   4   5   6   7
 2   3   4   5   6   7   8        8   9  10  11  12  13  14
 9  10  11  12  13  14  15       15  16  17  18  19  20  21
16  17  18  19  20  21  22       22  23  24  25  26  27  28
23  24  25  26  27  28  29       29  30  31
30
Month 5                          Month 11
Su  Mo  Tu  We  Th  Fr  Sa       Su  Mo  Tu  We  Th  Fr  Sa
     1   2   3   4   5   6                    1   2   3   4
 7   8   9  10  11  12  13        5   6   7   8   9  10  11
14  15  16  17  18  19  20       12  13  14  15  16  17  18
21  22  23  24  25  26  27       19  20  21  22  23  24  25
28  29  30  31                   26  27  28  29  30

Month 6                          Month 12
Su  Mo  Tu  We  Th  Fr  Sa       Su  Mo  Tu  We  Th  Fr  Sa
                 1   2   3                            1   2
 4   5   6   7   8   9  10        3   4   5   6   7   8   9
11  12  13  14  15  16  17       10  11  12  13  14  15  16
18  19  20  21  22  23  24       17  18  19  20  21  22  23
25  26  27  28  29  30           24  25  26  27  28  29  30
                                 31
*——————*——————*——————*——————*——————*——————*
```

Note: The calendar above is a jpeg image of the printout. Originally the author tried to copy the calendar in characters, but when the manuscript was processed for prining, the numbers became misaligned. Therefore, the only wasy to correctly illustrate the calendar in a good form was by jpeg image.

4.3 Interpolation of data in a table

A table of data is given as

x	y
1.200000	0.280725
1.400000	0.243009
1.600000	0.201810
1.800000	0.160976
2.000000	0.123060
2.200000	0.089584

Assume it is desired to find the values of y for x=1.35, 1.59 and 2.05 by interpolation. Interpolation may be performed by commad **interp1** with a few different options. The command **interp1 (x,y,xi)** is the default form of **interp1** that uses linear interpolation, where x and y are arrays of the data, and xi is the values of x for which the y-values are to be calculated by linear interpolation.

```
data=[
    1.200000    0.280725;
    1.400000    0.243009;
    1.600000    0.201810;
    1.800000    0.160976;
    2.000000    0.123060;
    2.200000    0.089584;
]
xi=[1.35   1.59   2.05]
interp1(data,xi)
ans =
    0.25244    0.20387    0.11469
```

Other options of interp1 include:
interp1(x,y,xi, 'cubic'): cubic Hermite which preserves the first derivative,
interp1(x,y,xi, 'spline'): spline interpolation with smooth first and second derivatives.

```
>>[x,y]= [
     1.200000    0.280725;
     1.400000    0.243009;
     1.600000    0.201810;
     1.800000    0.160976;
     2.000000    0.123060;
     2.200000    0.089584;
]
>>xi=[1.35  1.59  2.05]
>> interp1(x,y,xi,'spline')
ans =
     0.25293    0.20389    0.11422
```

4.4 Graph plotting for a function given as input

The program listed below plots a function. The function input is hard-coded in a string. The minimum and maximum of the axes of the graph and the number of points to express the curve are also hard-coded input.

```
%Function_plot
clf; clear all
disp('===================================================
=')

%INPUT
minmax = 'x_min=-2; x_max=3; y_min=-5; y_max=5;';
number_of_points=50;
Eq = 'y=sin(x)- 0.5*sin(2*x)';
%END OF INPUT

disp(['minmax of axes: ',minmax])
eval(minmax)
disp(['Equation to plot: ',Eq])
dx=(x_max-x_min)/Number_of_points;
x=x_min:dx:x_max;
y=eval(Eq);
plot(x,y)
Dx=(x_max-x_min); Dy=(y_max-y_min);
text(x_min+0.05*Dx, y_max-0.05*Dy, Eq,'fontsize',14)
axis([x_min,x_max,y_min,y_max])
xlabel('X','fontsize',16)
```

```
ylabel('X','fontsize',16)
disp('==================================================
=')
```
Output:
```
:==================================================
=:
```
minmax of axes: x_min=-2; x_max=3; y_min=-5; y_max=5;
Equation to plot: y=sin(x)- 0.5*sin(2*x)

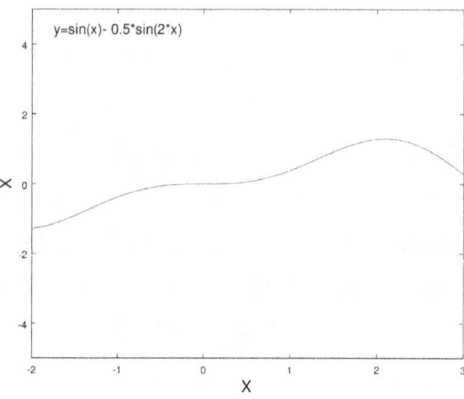

Figure 4.1 Plot of sin(x)-0.5*sin(2x)

4.5 Average and standard deviation

Standard Deviation is defined by

$$\sigma = \sqrt{\sum_{i=1}^{N}(x_i - \mu)^2 / N}$$

where N is the number of data, x_i is the i-th data, μ is the average of the data:

$$\mu = \sum_{i=1}^{N} x_i / N$$

Program:
```
%standard_deviation.m
disp('Input of data ')
x=[   26.806     60.041     55.406     62.741     28.978  51.007
62.101    35.569    25.266   31.689   ]
```

138

```
        disp('Average')
        mu=sum(x)/length(x)
        disp('Standard deviation')
        stdv= sqrt( sum( (x-mu).^2 )/length(x))
Output:
        Average
        mu =   43.960
        Standard deviation
        stdv =   14.867
```

4.6 Loan balance calculation

The method introduced here calculates the balance of loan with the following input data:

> Initial load amount, A
> Number of months of installemnt payments, $NoMo$
> Annual interest rate in percentage, R
> Monthly installemnt payment amount, $MoPay$

Assuming that the installment payment is done exactly on the last day of the month, the balance in the beginning of the month is calculated by

$$A=A*(1+R/100/12)-MoPay$$

where A on the right side is the balance of the previous month.

```
% loan_balance.m
%Loan balance calculator
% Suppose a person pays off his/her
% loan by monthly installemnt
% This program calculates the balance of
% the loan after n months of payments.
% original loand amount
disp('===========================================')
disp('Original loan amount, $')
A= 10000
disp('Annual interest rate')
R=5          % Annual interest rate in percent
IntRate=R/100;
disp('Maximum number of months to calculate')
NoMo=25      % Number of month
```

```
disp('Amount of fixed monthly payment')
MoPay=270
%
disp('================================================')
disp('Months=months after the load initiation ')
disp('Balance=load balance ')
disp('AccumInt=total interest that has been paid ')
disp('================================================')
m=0; %Initialization of number of months after load initiates
Ti=0; % Total interest initialized
for k=1:NoMo
m=m+1;
Ti=Ti+A*IntRate/12;
A=A*(1+IntRate/12)-MoPay;
if A>0,
fprintf('Months = %i, Balance=%4.2f, AccumInt=%4.2f \n', m,A,Ti)
end %if
if A<0, break; end
end %for
if A<0,
fprintf('Months = %i, Balance=%4.2f, AccumInt=%4.2f \n', m,A,Ti)
fprintf('Refund= %4.2f\n', -A)
end %if
disp('================================================')
disp(' ')
```

Output:

```
====================================================
Original loan amount
A =   10000
Annual interest rate
IntRate =   0.050000
Maximum number of months to calculate
NoMo =   25
Amount of fixed monthly payment
MoPay =   270
====================================================
Months=months after the load initiation
Balance=load balance
AccumInt=total interest that has been paid
====================================================
Months = 1, Balance=9771.67, AccumInt=41.67
Months = 2, Balance=9542.38, AccumInt=82.38
Months = 3, Balance=9312.14, AccumInt=122.14
```

Months = 4, Balance=9080.94, AccumInt=160.94
Months = 5, Balance=8848.78, AccumInt=198.78
Months = 6, Balance=8615.65, AccumInt=235.65
Months = 7, Balance=8381.55, AccumInt=271.55
Months = 8, Balance=8146.47, AccumInt=306.47
Months = 9, Balance=7910.41, AccumInt=340.41
Months = 10, Balance=7673.37, AccumInt=373.37
Months = 11, Balance=7435.35, AccumInt=405.35
Months = 12, Balance=7196.33, AccumInt=436.33
Months = 13, Balance=6956.31, AccumInt=466.31
Months = 14, Balance=6715.30, AccumInt=495.30
Months = 15, Balance=6473.28, AccumInt=523.28
Months = 16, Balance=6230.25, AccumInt=550.25
Months = 17, Balance=5986.21, AccumInt=576.21
Months = 18, Balance=5741.15, AccumInt=601.15
Months = 19, Balance=5495.07, AccumInt=625.07
Months = 20, Balance=5247.97, AccumInt=647.97
Months = 21, Balance=4999.84, AccumInt=669.84
Months = 22, Balance=4750.67, AccumInt=690.67
Months = 23, Balance=4500.46, AccumInt=710.46
Months = 24, Balance=4249.21, AccumInt=729.21
Months = 25, Balance=3996.92, AccumInt=746.92

4.7 Mortgage payment finder

The method here finds the exact amount of monthly installment payment for each month with given values of the initial mortgage loan amount, annual interest rate in percentage, number of years of loan. The following program uses a variation of Newton's iteration called the secant method. The final solution is the exact amount of monthly payment by which the mortgage balance becomes zero at the end of the mortgage period.

Program
```
%mortgage_payment.m
clear all
Str='for k=1:NoMo; A=A*(1+InMo)-MoPay; end %for';
disp('====== Mortgage monthly Payment Finder=======')
disp('Total amount of loan')
A0=150000
disp('Annual interest rate (percentage)')
```

```
InAn=5.5 % in PERCENT
disp('Number of years of mortgage')
NoYr=15
disp('. . . . . . .Iterative computations started . . . . . . .')
% Calculation starts
NoMo=NoYr*12;
InMo=InAn/100/12;
MoPay=A0/NoMo*.67;
fprintf('Initial guess for monthly payment=$%.2f', MoPay)

for itr=1:20
    if itr==2, MoPay=MoPay*2;
    fprintf('Second    guess for monthly payment=$%.2f', MoPay)
    end
    if itr>2, R=(bal(itr-1)-bal(itr-2) )/(mprec(itr-1)- mprec(itr-2) );
        MoPay=mprec(itr-1)- bal(itr-1)/R;
        fprintf('%i-th   guess   for   monthly payment=$%.2f', itr,MoPay)
    end %if
A=A0;
eval(Str)
%eval(S)
bal(itr)=A;
mprec(itr)=MoPay;
%[itr, MoPay, A]
fprintf('   Balance=$%0.2f\n', A)
if abs(A)<1 break; end
end
fprintf('Final Answer = $%0.2f per month\n', MoPay)
disp('====== END =======')
```

Output:

```
====== Mortgage monthly Payment Finder=======
Total amount of loan
A0 =    150000
Annual interest rate (percentage)
InAn =   5.5000
Number of years of mortgage
NoYr =    15
. . . . . . .Iterative computations started . . . . . . .
Initial guess for monthly payment=$558.33   Balance=$186004.63
Second   guess for monthly payment=$1116.67   Balance=$30371.70
3-th   guess   for   monthly payment=$1225.63   Balance=$0.00
Final Answer = $1225.63 per month
====== END =======
```

4.8 Dice game

In the present game, two dices are rolled. There is no win/fail in the game, but such a rule may be built in by the reader. For example, two players play alternatively. If two dices show the same number, the person who rolled the dices may get 5 cents. The program may count the total gain for each player. Feel free to change the program in any way desired. Have a fun.

Program:
```
%dice_game.m
clear all, clf
in=1;kount=0
while in<998
kount=kount+1; % kount=Game count
disp('If you want continue game, type 1 and hit return.')
in=input('To stop game, type 999 and hit return')
c = ceil(6*rand(1,2));
a=c(1)%a is the reading of the left dice
b=c(2)%b is the reading of the right dice
clf
axis([2 10 0 8]); axis off;hold on
%1
  kt=num2str(kount)
text(3,7,['Game count=',kt], 'fontsize', 14)
if a==1,
plot(4.5,4.5,'*', 'linewidth',40)
elseif a==2'
plot([4.2,4.2],[ 4.2,4.8],'*',            'linewidth',20);
elseif a==3,
plot([4.0,4.5, 5.0],[4, 4.5, 5.], '*', 'linewidth',20);
elseif a==4,
plot([4   5 4 5],[4 4 5 5], '*', 'linewidth',20);
elseif a==5,
plot([4   5 4.5 4 5],[4 4 4.5 5 5], '*', 'linewidth',20);
elseif a==6,
plot([4   5 4 5 4 5],[4 4 4.5 4.5 5 5], '*', 'linewidth',20);
end %if
if b==1,
```

```
plot(4.5+3.5,4.5,'*', 'linewidth',40)
elseif b==2'
plot([4.2,4.2]+3.5,[ 4.2,4.8],'*',           'linewidth',20);
elseif b==3,
plot([4.0,4.5, 5.0]+3.5,[4, 4.5, 5.], '*', 'linewidth',20);
elseif b==4,
plot([4   5 4 5]+3.5,[4 4 5 5], '*', 'linewidth',20);
elseif b==5,
plot([4   5 4.5 4 5]+3.5,[4 4 4.5 5 5], '*', 'linewidth',20);
elseif b==6,
plot([4   5 4 5 4 5]+3.5,[4 4 4.5 4.5 5 5], '*', 'linewidth',20);
end %if
end %while
clf
```
Output: Please play and see.

4.9 Sierpinski triangle

Sierpinski Triangle is a fractal consisting of triangles of different sizes. However, every triangle has the same shape regardless to the size. There are different ways of drawing the Sierpinski triangle, among which two are introduced next.

Method 1 : First, draw apices and sides of the largest equilateral triangle using a ruler. Mark the mid points of the three sides. By connecting the mid points, divide the largest triangle into four smaller triangles. The triangle at the center among the four will be left untouched as a white triangle. Each of the three remaining triangles shares one of the apices of the largest triangle. Repeat the same operations performed to the largest triangle to those three smaller triangles except the center triangle. As the triangles become smaller and smaller, keep repeating the same operations until the triangles become unworkably small.

Figure 4.2   Siepinski Triangle

Method 2 : Prepare a dice, a sheet of paper, a compass to find the midpoint of a line, a sharp pencil, and two plastic triangles. As in Method 1, draw the largest equilateral triangle.   Name the apices as p1, p2, and p3.

Find the center point between the top apex and the center point of the bottom side of the largest triangle.   Mark the point by a dot named c0. Choose one from p1, p2, and p3 using the dice (get p1 if the dice is 1 or 2, p2, if the dice is 3 or 4, and p3 if the dice is 5 or 6).

Find the center point between c0 and p1, p2 or p3 whichever has been chosen. The center point is marked by a dot named c1. Choose one from p1, p2 and p3 using the dice, and find the center points between c1 and p1, p2 or p3 whichever has been chosen. Mark it by a dot and name as c2. Repeat the same procedure over and over.   A rough image of the fractal becomes clear after marking a few hundred dots.

The figure above was created by a program on Octave that emulates Method 2 using random numbers in place of a dice. The number of the dots marked is 25,000.

Program:
```
%sierpinski.m
p(1,:)=[0,0];
p(2,:)=[1,0];
p(3,:)=[0.5,sqrt(0.75)];
randi=0.1
```

```
z=[0.5,0.0]
hold off
clf,   axis square,   hold on
for n=2:25000
      k=ceil(rand*3);
      z=0.5*(z + p(k,:));
      plot([z(1), z(1)+0.001], [z(2),z(2)], 'linewidth',0.5)
end
```
Output: a graph shown above.

Method 2 is a very simple algorithm, yet it is inexplicably mysterious why white triangles are left untouched by the dots.

4.10 Apollonian gasket

How to construct Apollonian gasket
In Figure 4.3 two small circiles inside that large circile are touching each other. In Figure 4.4 an additional circile is drawn touching other two circles. The question we ask here is how to draw a circle that touches other two circles plus the outer circle, because this is the fundamental technique necessary to construct an Applonian gasket.

Figure 4.3   Figure 4.4

There are two ways to answer the question. The first is to use only a compass and a ruler, and the second is to use a mathematical method to find the coordinates of the intersection of two circles.

Geometrical approach

Let us call the largest circle as c_0 in Figure 1. We also name the two circles inside c_0 as c_1 and c_2. The radii of c_1 and c_2 are denoted by r_1 and r_2 respectively.

The new circle in Figure 4.4 is c_3. The radius of c_3 namely r_3 is unknown at this time, but we make an arbitrary estimate.

Draw a circle with radius r_1+r_3 sharing the center with c_1. Draw another circle with radius r_2+r_3 sharing the center with c_2.

If we draw a circle of radius r_3 around the center of the intersection between the two circles drawn in accordance with the prior paragraph, naming it as c_3, c_3 is touching tangentially with c_1 and c_2, but it does not touch c_0 except by accident. Therefore we need to change r_3 so c_3 will touch c_0.

Draw a straight line connecting the centers of c_3 and c_0 stretching to the circle c_0, where it intersect both c_0 and c_3 at the right angle. Get the half length between the intersections of the line with c_3 and c_0. Add it to r_3, and we will consider it as a revised r_3.

Repeat the procedure written in the third through the fifth paragraphs. The value of revised r_3 will converge after a few iterations.

Analytical approach
Although the analytical approach we write here is essentially the same as the geometrical approach, the coordinates of the intersection between the circle of radius r_1+r_3 and that with r_2+r_3 are obtained analytically.

Let (x_1,y_1) and (x_2,y_2) be the coordinates of the center of c_1 and c_2 respectively. The equation that represent the circles of radius r_1+r_3 and that r_2+r_3 are respectively

$$(x - x_1)^2 + (y - y_1)^2 = (r_1+r_3)^2$$

$$(x - x2)^2 + (y - y2)^2 = (r2+r3)^2$$

The coordinates of the intersection between these circles may be obtained by Newton's iteration (see Section 3.4). Let $(x_0 + \delta x, y_0 + \delta y)$ be the coordinates of the intersection, where (x_0, y_0) are estimates for the coordinates for the center, and δx and δy are corrections. We consider $(x_0 + \delta x, y_0 + \delta y)$ as the correct coordinates, for which the original equations become

$$(x_0 + \delta x - x1)^2 + (y_0 + \delta y - y1)^2 = (r1+r3)^2$$

$$(x_0 + \delta x - x2)^2 + (y_0 + \delta y - y2)^2 = (r2+r3)^2$$

We expand them and ignore δx^2 and δy^2 to get a set of linear equations for δx and δy as

$$2(x_0 - x1)\,\delta x + 2(x_0 - y1)\,\delta y = (r1+r3)^2 - (x_0 - x1)^2 - (y_0 - y1)^2$$

$$2(x_0 - x2)\,\delta x + 2(y_0 - y2)\,\delta y = (r2+r3)^2 - (x_0 - x2)^2 - (y_0 - y2)^2$$

By solving the foregoing equations we get $(x_0 + \delta x,\ y_0 + \delta y)$ but they are not exact coordinates of the intersection because we ignored δx^2 and δy^2. Therefore, using $(x_0 + \delta x,\ y_0 + \delta y)$ as revised estimates for the coordinates, we repeat the same procedure a few more times. Convergence of iteration is normally very fast and the coordinates become accurate to 7~8 digits within a few iterations. The only thing we need to be careful about is that there are two intersections between two circles, so unless the estimated coordinates are closer to those of the desired intersection we may get wrong answer.

If the operations of adding circles touching three circles is continued. Figure 4.5 below is reached.

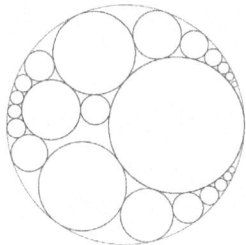

Figure 4.5

Figure 4.5 would become different if the size and location of the circles in Figure 4.3 are changed. The effort on Figure 4.5 may be still continued with smaller and smaller circles, but some change of the rule would be necessary, that is, in Figure 4.3 and 4.4, we ruled that the added circles must touch the outer circle tangentially. However, in order to fill the spaces available with smaller circles, the new circles need to touch only any three circles. At any rate, a lot of patient work will be necessary to add more circles.

Figure 4.5 was drawn as an author's casual trial, but it is related to the fractals called Apollonian Gasket created by an ancient Greek mathematician Apollonius.

Apollonian gasket drawing
Apollonian gaskets include many beautiful fractals with different configurations of symmetries. Only one of them is illustrated in Figure 4.6. Figure 4.6 was plotted by an author's program, but lacks infinitely small triangles because the author's program is not fully automated. So it is a unfinished Apollonian Gasket.

The source code to plot Figure 4 is available for copy/paste in http://octave.ismr.us/quiz/more.htm

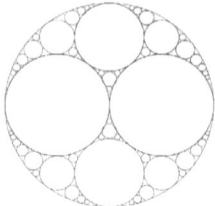

Figure 4.6   Apollonian Gasket Example

```
%apollo2.m
clf;
figure(1);
x0=0; y0=0; r0=4;
w0(1)=x0; w0(2)=y0;, w0(3)=r0;
circle_(x0,y0,r0);hold on
r3_initial=1;
x1=2; y1=0; r1=2;
w1(1)=x1; w1(2)=y1;, w1(3)=r1;
circle_(x1,y1,r1)
w2(1)=-2; w2(2)=y1; w2(3)=r1;
circle_(w2(1),w2(2),w2(3))
r3_initial=2
w3=tan_cir(w0,w1,w2, r3_initial,0,2)
circle_(w3(1),-w3(2),w3(3))
zs1=space_3c(w3,w1,w2, 0.1, 0,1)
zs11=space_3ctest(zs1,w1,w2, 0.002, 0, 0.6)
zs13=space_3ctest(zs1,w1,w3, 0.002, 0.1, 0.1)
zs23=space_3ctest(zs1,w2,w3, 0.002, -0.1, 0.1)
circle_(zs1(1),-zs1(2),zs1(3))
circle_(zs13(1),-zs13(2),zs13(3))
circle_(zs11(1),-zs11(2),zs11(3))
circle_(zs23(1),-zs23(2),zs23(3))
w134=space_3ctest(w4,w1,w3, 0.1, 1.7, 2)
circle_(w134(1),-w134(2),w134(3))
circle_(-w134(1),w134(2),w134(3))
circle_(-w134(1),-w134(2),w134(3))
w4=tan_cir(w0,w1,w3, 1,2,2.5)
w34=tan_cir(w0,w4,w3, 0.5,1.5,3.2)
circle_(w34(1),-w34(2),w34(3))
circle_(w4(1),-w4(2),w4(3))
circle_(-w34(1),-w34(2),w34(3))
circle_(-w4(1),-w4(2),w4(3))
circle_(-w34(1),w34(2),w34(3))
circle_(-w4(1),w4(2),w4(3))
w1341=tan_cir(w0,w34,w3, 0.25,1.,3.
circle_(w1341(1),-w1341(2),w1341(3))
circle_(-w1341(1),w1341(2),w1341(3))
circle_(-w1341(1),-w1341(2),w1341(3))
w13411=tan_cir(w0,w1341,w3, 0.15,0.7,3.8)
circle_(w13411(1),-w13411(2),w13411(3))
circle_(-w13411(1),w13411(2),w13411(3))
circle_(-w13411(1),-w13411(2),w13411(3))
w5=tan_cir(w0,w1,w4, 1,3,2)
w45=tan_cir(w0,w5,w4, 0.4,2.5,2.6)
circle_(w5(1),-w5(2),w5(3))
circle_(w45(1),-w45(2),w45(3))
circle_(-w5(1),-w5(2),w5(3))
```

```
circle_(-w45(1),-w45(2),w45(3))
circle_(-w5(1),w5(2),w5(3))
circle_(-w45(1),w45(2),w45(3))
w6=tan_cir(w0,w1,w5, 0.7,3.5,2)
circle_(w6(1),-w6(2),w6(3))
circle_(-w6(1),-w6(2),w6(3))
circle_(-w6(1),w6(2),w6(3))
w7=tan_cir(w0,w1,w6, 0.4,3.8,1.8)
circle_(w7(1),-w7(2),w7(3))
circle_(-w7(1),-w7(2),w7(3))
circle_(-w7(1),w7(2),w7(3))
w8=tan_cir(w0,w1,w7, 0.2,3.8,1.6)
circle_(w8(1),-w8(2),w8(3))
circle_(-w8(1),-w8(2),w8(3))
circle_(-w8(1),w8(2),w8(3))
w9=tan_cir(w0,w1,w8, 0.15,3.9,1.2)
circle_(w9(1),-w9(2),w9(3))
circle_(-w9(1),-w9(2),w9(3))
circle_(-w9(1),-w9(2),w9(3))

%%%%%%%%%%%%%%%%%%%%%%%%%%%%%%%%
function a=circle_(xc, yc, r)
%xc=1;yc=1;r=1;
dth=pi/500;
th=0:dth:pi*2;
x=r*cos(th)+ xc;
y=r*sin(th)+ yc;
plot(x,y);
axis square
%%%%%%%%%%%%%%%%%%%%%%%%%%%%%%%%%
function w=tan_cir(w0,w1,w2, r3_initial, xg,yg)
x0=w0(1); y0=w0(2); r0=w0(3)
for m=1:5
x1=w1(1); y1=w1(2); r1=w1(3); r3=r3_initial;
circle_(x1,y1,r1);
%circle_dot(x1,y1,r1+r3)
x2=w2(1); y2=w2(2); r2=w2(3);
circle_(x2,y2,r2)
%circle_dot(x2,y2,r2+r3)
axis([-3 8 -5 5])
x=xg; y=yg;
for itr=1:5
r1d=r1+r3; r2d=r2+r3;
a(1,1)=2*(x-x1); a(1,2)=2*(y-y1); b(1)=r1d^2 - (x-x1)^2 - (y-y1)^2;
a(2,1)=2*(x-x2); a(2,2)=2*(y-y2); b(2)=r2d^2- (x-x2)^2 - (y-y2)^2;
s=a\b';
x=x+s(1);
y=y+s(2);
%[x,y,itr]
end
%line([x,x0], [y,y0])
EL=sqrt( (x-x0)^2 + (y-y0)^2);
r3n=0.5*r3+0.5*( r0-EL);
r3_initial=r3n;
%fprintf('[r0,EL,r3,r3n, r0-EL]');[r0,EL,r3,r3n, r0-EL]
end
w(1)=x; w(2)=y; w(3)=r3n;%[w, 999]
circle_(x,y,r3)
```

```
%%%%%%%%%%%%%%%%%%%%%%%%%%%%%%%%%%%%
function w=space_3ctest(w0,w1,w2, r3_initial, xg,yg)
x0=w0(1); y0=w0(2); r0=w0(3)
for m=1:10
x1=w1(1); y1=w1(2); r1=w1(3); r3=r3_initial;
circle_(x1,y1,r1);
%circle_dot(x1,y1,r1+r3)
x2=w2(1); y2=w2(2); r2=w2(3);
circle_(x2,y2,r2)
%circle_dot(x2,y2,r2+r3)
axis([-3 8 -5 5])
x=xg; y=yg;
for itr=1:5
r1d=r1+r3; r2d=r2+r3;
a(1,1)=2*(x-x1); a(1,2)=2*(y-y1); b(1)=r1d^2 - (x-x1)^2 - (y-y1)^2;
a(2,1)=2*(x-x2); a(2,2)=2*(y-y2); b(2)=r2d^2- (x-x2)^2 - (y-y2)^2;
s=a\b';
x=x+s(1);
y=y+s(2);
%[x,y,itr]
end
%line([x,x0], [y,y0])
EL=sqrt( (x-x0)^2 + (y-y0)^2);
r3n=r3+0.25*( EL-r0-r3);
r3_initial=r3n;
%fprintf('[r0,EL,r3,r3n, r0-EL]');[r0,EL,r3,r3n, EL-r0]
end
w(1)=x; w(2)=y; w(3)=r3n;%[w, 999]
circle_(x,y,r3n)


%%%%%%%%%%%%%%%%%%%%%%%%%%%%%%%%%%%%%
function w=space_3c(w0,w1,w2, r3_initial, xg,yg)
x0=w0(1); y0=w0(2); r0=w0(3)
for m=1:5
x1=w1(1); y1=w1(2); r1=w1(3); r3=r3_initial;
circle_(x1,y1,r1);
%circle_dot(x1,y1,r1+r3)
x2=w2(1); y2=w2(2); r2=w2(3);
circle_(x2,y2,r2)
%circle_dot(x2,y2,r2+r3)
axis([-3 8 -5 5])
x=xg; y=yg;
for itr=1:5
r1d=r1+r3; r2d=r2+r3;
a(1,1)=2*(x-x1); a(1,2)=2*(y-y1); b(1)=r1d^2 - (x-x1)^2 - (y-y1)^2;
a(2,1)=2*(x-x2); a(2,2)=2*(y-y2); b(2)=r2d^2- (x-x2)^2 - (y-y2)^2;
s=a\b';
x=x+s(1);
y=y+s(2);
%[x,y,itr]
end
%line([x,x0], [y,y0])
EL=sqrt( (x-x0)^2 + (y-y0)^2);
r3n=0.5*r3+0.5*( EL-r0);
r3_initial=r3n;
fprintf('[r0,EL,r3,r3n, r0-EL]');[r0,EL,r3,r3n, r0-EL]
end
w(1)=x; w(2)=y; w(3)=r3n;%[w, 999]
circle_(x,y,r3n)
```

4.11 Sudoku solution

Sudoku quizes are very popular and can be found in magazines and newspapers. They are nice quizes if tons of spare time is available. However, if you are challenged by someone like a friend, one of your children or grand children on solving-sudoku competition, here is a secret: use Octave/Matlab.

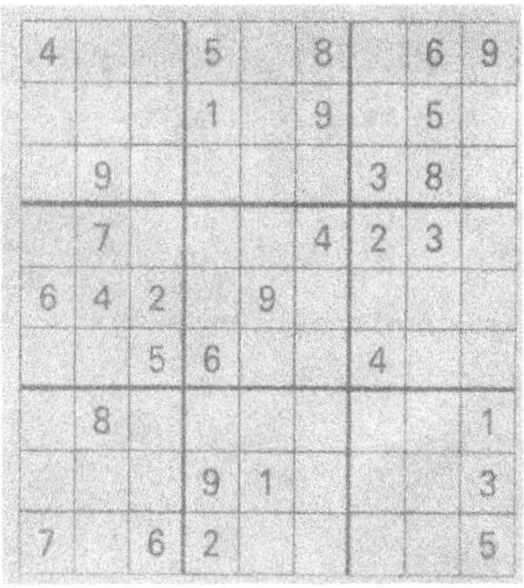

Figure 4.7 Example of Sudoku

Algorithm of solution is as follows. First, numbers eligible for each vacant slot are searched and recorded. Second, if there are vacancies that have only one eligible number is found, fill the vacancies with the found numbers. These two steps are repeated until all the vacancies are filled.

If there is no vacancy that has only one eligible number, that sudoku problem must be at a higher level and need many trials and errors. That is, one vacancy is filled with one of the eligible numbers, and the two steps are repeated. If any conflict is encountered, the choice of the eligible number is not correct one, so the next eligible number is tried. If all the eligible numbers fail, then move to the next vacancy and try with all eligible number, and so on. It will take a lot of time and effort. Normally, the sudoku problems in the newspaper or magazine is not in this high level, so they can be solved by the method written in the prior paragraph.

The program listed below is only for the low level. It solved all the sudoku problems author found in the newspaper.

The source code below is also available for copy/paste in http://octave.ismr.us/sudoku.htm. The only place to alter in the program is 9-by-9 matrix of the sudoku problem. Use 0 for blank entries in the sudoku problem. The answer will come in a blink of time.

```
>> sudoku_test002
Problem
```

4	0	0	5	0	8	0	6	9
0	0	0	1	0	9	0	5	0
0	9	0	0	0	0	3	8	0
0	7	0	0	0	4	2	3	0
6	4	2	0	9	0	0	0	0
0	0	5	6	0	0	4	0	0
0	8	0	0	0	0	0	0	1
0	0	0	9	1	0	0	0	3
7	0	6	2	0	0	0	0	5

```
Solution
```

4	2	7	5	3	8	1	6	9
3	6	8	1	4	9	7	5	2
5	9	1	7	6	2	3	8	4
1	7	9	8	5	4	2	3	6
6	4	2	3	9	7	5	1	8
8	3	5	6	2	1	4	9	7
9	8	3	4	7	5	6	2	1
2	5	4	9	1	6	8	7	3
7	1	6	2	8	3	9	4	5

Main program

```
%cccccccccccccccccccccccccccccccccccccccccccccccccccc
clear all
disp('Sudoku Problem')
a=[
4 0 0 5 0 8 0 6 9;
0 0 0 1 0 9 0 5 0;
0 9 0 0 0 0 3 8 0;
0 7 0 0 0 4 2 3 0;
6 4 2 0 9 0 0 0 0;
0 0 5 6 0 0 4 0 0;
0 8 0 0 0 0 0 0 1;
0 0 0 9 1 0 0 0 3;
```

7 0 6 2 0 0 0 0 5
]Level=1;
[aa, ir, nee,iok,a0total]=makeir(a,Level);
disp('Sudoku Solution')
a=aa

Function makeir.m

```
%cccccccccccccccccccccccccccccccccccccccccccccccccccc
% Finds compatible numbers for each entries of a
% If vacant entry of a has only one candidate
% it fills it after checking of no conflicts.
% The above procedure is repeatedd iteratively
% until no single candidates can be found.
function [a_, ir, nee,iok,a0total]=makeir(a,Level)
%
for iter=1:50
[nee, ir]=sub0(a);
Ltotal=0; a0total=0; nenz=0;
for i=1:9
for j=1:9
net=nee(i,j);%[ i,j, ne(i,j), ir(i,j,1:net)]
if (Level==1 & iter<=2),
%[ i,j, ne(i,j), ir(i,j,1:net)]
end %if
Ltotal=Ltotal+nee(i,j);
if (a(i,j)==0) a0total=a0total+1; end %if
if (nee(i,j) ~= 0) nenz=nenz+1; end %if
end %for
end %for
%          [iter, a0total, nenz]
norep=0;
for i=1:9
for j=1:9
net=nee(i,j);
%c  If only one num is available, it is enterted in a(i,j)
if (net==1),
icompt= compati(a,i,j,ir(i,j,1));
if (icompt==1),
a(i,j)=ir(i,j,1);
norep=norep+1;
ir(i,j,1)=0;
nee(i,j)=0;
end %if
end %if
end %for
```

```matlab
end %for
%    if (Level==1 & iter<=2), a, end %if
if (norep==0) break; end %if
end %for % terminates iteration, iter
if (a0total>nenz),
disp('current setting is imcompatible')
iok=0; a_=a; return
elseif (a0total==nenz),
iok=1;
% fprintf('currnt setting is OK, a0total=%i   ...
%    nenz=%i\n', a0total, nenz)
end %if
a_=a;
return
```

Function sub0.m

```matlab
%cccccccccccccccccccccccccccccccccccccccccccccccccc
%    This subroutine actually develops ir
%    Finds compatible numbers for vacant entries in a
function [nee,ir]=sub0(a)
nee(1:9,1:9)=0;
ir(1:9,1:9,1:9)=0;
for i=1:9
for j=1:9
if (a(i,j)==0),
for L=1:9
icompt=compati(a,i,j,L);
if (icompt==1),
nee(i,j)=nee(i,j)+1;
net=nee(i,j);
ir(i,j,net)=L;
end %if
end %for        %L loop
end %if      % if (a(i,j)==0)
end %for
end %for
return
```

Function compat.m

```matlab
%cccccccccccccccccccccccccccccccccccccccccccccccccc
%    Checks if L is compatible at i and j;
%     if yes, compat=1, if not compat=0
function compat=compati(a,i,j,L)
if (i>=1 & i<=3), k1=1; k2=3; end %if
if (i>=4 & i<=6), k1=4; k2=6; end %if
```

```
if (i>=7 & i<=9), k1=7; k2=9; end %if
if (j>=1 & j<=3), m1=1; m2=3; end %if
if (j>=4 & j<=6), m1=4; m2=6; end %if
if (j>=7 & j<=9), m1=7; m2=9; end %if
% Examining the 3x3 box for compatibility
for i9=k1:k2
for j9=m1:m2
if (L==a(i9,j9)) compat=0; return ; end %if
end %for
end %for
%c   Examining the row i for compatibility
for j8=1:9
if (L==a(i,j8)) compat=0; return ; end %if
end %for
%c Examining the column for compatibility
for i8=1:9
if (L==a(i8,j)) compat=0; return ; end %if
end %for
%c       L passed eligibility for box (i,j)
compat=1;
return
```

4.12 Self-intersecting surfaces

Figure 4.8 illustrates the surface of a closed space created by the following equations:

$$x(u, v)=(1+\cos(v))\cos(u)$$
$$y(u, v)=(1+\cos(v))\sin(u) \qquad (1)$$
$$z(u, v)=-\tanh(u -\pi)\sin(v)$$

where u and v change from 0 to 2π. The surface is a self-intersecting surface [Reference: Wikipedia]. The shell may be opened as shown in Figure 4.9 by changing the third equation above to

$$z(u, v)=-\tanh(u -\pi)\sin((y-\pi)/1.3+\pi) \qquad (2)$$

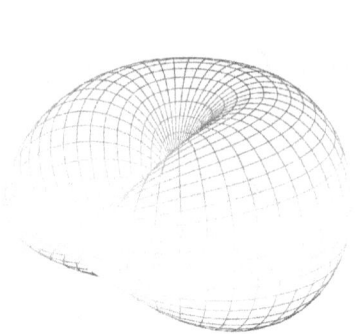

Figure 4.8 Equation 1   Figure 4.9 Equation 2

```
%Plot of Figure 4.8 and 4.9
clear all, clf
h=pi/40;
u=0:h:2*pi;
v=0:1/40:1;
[x,y]=meshgrid(u,v);
while 1
X=y.*cos(2*x);
Y=y.*sin(2*x);
c=input('-1 closed, 2 power of input; 0 quit   ')
if c==-1,
Z=sin(pi*y).*cos(x);
elseif c>0;
Z=(y.^c).*cos(x);
else c==0;   return
end
mesh(X,Y,Z)
axis off
%if c==2; axis square;end
end
```

Figure 4.10 illustrates plot of another set of equations given by

$$x(u, v) = v \cos(2u)$$
$$y(u, v) = v \sin(2u) \qquad (3)$$
$$z(u, v) = \sin(\pi v) \cos(u)$$

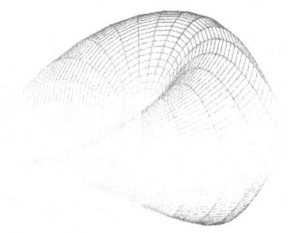

Figure 4.10 Shell with Equation (3)

The third line of Eq.(3) is changed to

$$z(u, v) = v^c \cos(u) \qquad (4)$$

where c is the power. Figures 4.11 and 4.13 are images of opened shells using Eq.(4):

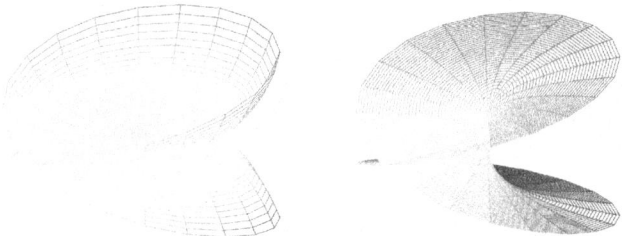

Figure 4.11 Eq.(4) with c=4   Figure 4.12 Eq.(4) with c=0.3

```
%Plot of Figures 4.10-4.12
clear all, clf
h=pi/20;
u=0:h:2*pi;
v=0:1/40:1;
[x,y]=meshgrid(u,v);
while 1
X=y.*cos(2*x);
Y=y.*sin(2*x);
c=input('-1 closed, 2 power of input; 0 quit   ')
if c==-1,
Z=sin(pi*y).*cos(x);
elseif c>0;
Z=(y.^c).*cos(x);
else c==0;   return
end
mesh(X,Y,Z)
```

```
axis off
%if c==2; axis square;end
end
```

Exercise problems for Chapter 4

[0] Oops! No, there is no exercise problems fabricated by the author. Make your own challenges and enjoy.

Appendices

Appendix 1
User-defined mathematical functions
(Use of eval and feval explained)

Most primitive way of using mathematical functions is to write the function directly in the main script. In this way, however, the same function may have to be written over and over if the function is used multiple times in the whole script. However, there are many convenient ways of defining and using it:

(1) Define a function as a separate function and save it as a m-file

Define a function m-file, for example

```
function f=my_f(x)
f=sin(2*x)+x.*2;
```

and saved in the same directory as the script calling it. Do not forget to use array arithmetic operators such as .^,   .* and ./ in the function m-file.

Then, it can be used in a script:

```
clear
x=0:0.1:3*pi;
f=my_f(x);
plot(x,f)
```

Alternatively, feval may be used:

```
clear
x=0:0.1:3*pi;
f=feval('my_f', x);
plot(x,f)
```

(2) A function name may be defined as string in the beginning of a program, and then **eval** is used whenever using the function is necessary

```
clear
x=0:0.1:3*pi;
s='my_f(x)';
f=eval(s);
plot(x,f)
```

(3) The function name may be read as input:

```
clear
x=0:0.1:3*pi;
%s='my_f(x)';
s=input(' Input your function name:   ')
f=eval(s);
plot(x,f)
```

Here, my_f must be enclosed by single quote signs as:

```
Input your function name:    'my_f'
s = my_f
```

By adding "s" at the end of **input command,** the input does not have to be enclosed by single quote signs:

```
clear
x=0:0.1:3*pi;
s=input(' Input your function name:   ',"s")
f=feval(s,x);
plot(x,f)
```

Input procedure:
```
Input your function name:   my_f
s = my_f
```

However, "s" works only in Octave, but not in Matlab.

(4) Methods of defining the function without function m-file

The definition of an equation can be inputted without using function m-file. In the following script, the equation is inputted through the keyboard (not enclosing by single quote signs):

```
clear
x=0:0.1:3*pi;
s=input(' Input your function name:   ',"s")
f=eval(s);
plot(x,f)
```

Input procedure:
```
Input your function name:   f=sin(2*x)+x.*2
s = f=sin(2*x)+x.*2
```

Here, s=f=... may be confusing, but s is the sting name and f=sin(2*x)+x.*2 is the string.

As already written, "s" works only in Octave, but not in Matlab.

Appendix 2
Examples of using eval command

In this appendix, item (4) in Appendix 1 is explained in more details. We show how to make writing a program that uses the same mathematical function many times easy. To explain how, consider the script below to solve the nonlinear equation,

$$\log(1+x) - x^2/10 = 0$$

by Newton's iteration (see Section 3.4):

```
% Appendix 2 Without using eval
clf,clear
x4plot=0:0.1:10; xinitial=5;
dx=0.001; x=xinitial;
for k=1:20
z=x; f=log(1+x)-x.^2/10; plot(x,f,'*'); hold on
xp=x+dx; fp=log(1+xp)-xp.^2/10;
fd=(fp-f)/dx; x=x-f/fd;; if abs(f/fd)<0.000001, break; end %if
end; residue=f/fd;
```

```
fprintf(['Equation="','log(1+z)-z.^2/10', '=0": Solution=%.5f\n'], x)
fprintf('Total iteration=%i, Residue=%.3e \n', k, residue)
x=x4plot; z=x; f=log(1+z)-z.^2/10;
plot(x,f)
```

The same equation is written several times in the script, so adapting it to another function needs much work. In order to make it easier to change, it is desired to write the equation only once in the entire script.

This is achieved by defining the function as a string variable, and using it by the **eval** command. In the following revised script, the function is written only once to define as string variable **fn**. Whenever the function is used, it is evaluated by **eval(fn)** with an appropriate definition of the argument just before **eval(fn)** is used. When the script is altered for another equation, only the second line needs to be changed, although the initial guess and **x4plot** that defines the plotting parameters may have to be adjusted also. If the revised script below is saved under a proper name, it is much more convenient to apply to many different equations than the previous one.

```
% Appendix 2 Using eval
clf,clear
fn='log(1+z)-z.^2/10';
x4plot='0:0.1:10'; xinitial=5;
dx=0.001; x=xinitial;
for k=1:20
z=x; f=eval(fn); plot(x,f,'*'); hold on
xp=x+dx;
fp=log(1+xp)-xp.^2/10;
z=xp; fp=eval(fn);
fd=(fp-f)/dx; x=x-f/fd; if abs(f/fd)<0.000001, break; end %if
end;
residue=f/fd;
fprintf(['Equation="',fn, '=0": Solution=%.5f\n'], x)
fprintf('Total iteration=%i, Residue=%.3e \n', k, residue)
x=eval(x4plot); z=x; f=eval(fn);
plot(x,f)
```

Appendix 3

How to make MyOctavePlace a default directory

Let us assume that Octave was downloaded and installed very recently, so the home directory where Octave opens by default is now

c:/users/owner/

or an equivalent name because the name of directory **owner** may be different on each different computer. In this writing we assume **owner** is the current default directory name when Octave is installed. The reader must understand this and adjust the name if the current default directory name is different. We also assume that a directory **MyOctavePlace** was created in **c:/users/owner/** with an intention to make it Home directory for Octave works.

We now wish to make

c:/users/owner/MyOctavePlace

the Home directly so Octave will open automatically in this directory. Otherwise, working directory must be changed manually from **c:/users/owner/** to **c:/users/owner/MyOctave Place** by using **>>cd MyOctavePlace**.

The following is the procedure in GUI Command Window. Copy the following two lines to the command window and run:

>>cd c:/Octave/octave-4.0.0/share/octave/site/m/startup/
>>edit octaverc

Then, the file octaverc opens in the Editor. Add the following line at the end:

setenv('HOME', 'C:/users/owner/MyOctavePlace/'); cd ~/

Save the octaverc file. Close Octave and restart the computer.

If everything goes well with no errors, the Octave should open in MyOctavePlace directory as the Home directory. Make sure >>**pwd** is responded by

c:/users/owner/MyOctavePlace

Appendix 4
Octave can become fatally sick

The readers should know **Octave can become fatally sick** if it swallows a wrong command or variable name. Here is the author's embarrassing experience.

At one time the author was writing a short m-file. During the first test, Octave stopped working properly, but there was no obvious error found in the m-file. However, whenever the command **plot** is used, error messages were generated. So the author tested the **plot** command independently of the m-file, but the result was the same. Other programs using **plot** stopped working in the same way.

Such ill behavior continued after Octave was shut down and re-opened, and even after the computer was shut down and rebooted. Another strange thing was that the m-file never ran even after almost all the contents including **plot** command were removed. It seemed as if Octave never forget that particular name of the m-file and continued attacking cantankerously.

Finally the only way for remedy was thought to re-install the whole Octave using the installer file downloaded.

It turned out later that the most probable cause was that the variable name **linewidth** was used accidentally as user-defined variable in the m-file tested. This name must be a reserved name used as a parameter in some graphic commands. My guess is that the "user-defined **linewidth**" permanently destroyed the Octave system, although **>>exist linewidth** returns **ans=0**.

After this incidence, though, Octave has been friendly to the author.

Matlab can possibly get similar symptom. The author has not attempted to cause the same trouble, nor is willing to try primarily because reinstallation of Matlab is not so simple to the author.

Solutions of Problems

Exercise problems for Chapter 1

[1]

```
% Problem chapter 1 No.1
v=rand(1,100);
avarge=sum(v)/length(v);
n=0;m=0;
for i=1:length(v)
if v(i)>0.5, n=n+1; end
if v(i)<0.5, m=m+1; end
end %for
fprintf('Number of entries greater than 0.5 is %i \n',n)
fprintf('Number of entries less than 0.5 is %i \n',m)
```

[2]

```
% Problem Chapter 1 [2] (i)
clear all
 x=[-5, -2, 0, 3, 5];
for i=1:length(x)
 y(i) = x(i)^2 - 2*x(i) -2 ;
  fprintf('x= %.3f    y=%.3f \n', x(i), y(i))
end %for
```

```
x= -5.000      y=33.000
x= -2.000      y=6.000
x=   0.000     y=-2.000
x=   3.000     y=1.000
x=   5.000     y=13.000
```

```
% Problem Chapter 1 [2] (ii)
clear all
 x=[-5, -2, 0, 3, 5];
 y=x.^2 - 2*x -2 ;
for i=1:length(x)
fprintf('x= %.3f    y=%.3f \n', x(i), y(i))
end %for
```

[3]

```
% Problem Chapter 1 [3]
clear all
disp('Next input must be enclosed by single quotes');
s=input('Input the equation in a string: ');
disp('Next input must be placed in [   ] ');
```

```
x=input('Input the x values in an array: ');
eval(s)
for i=1:length(x)
fprintf('x= %.3f    y=%.3f \n', x(i), y(i))
end %for
```

Next input must be enclosed by single quotes
Input the equation in a string: 'y=x.^2 - 2*x -2'
Next input must be placed in [   ]
Input the x values in an array: [-5, -2, 0, 3, 5]
y =
     33   6   -2   1   13

```
x= -5.000      y=33.000
x= -2.000      y=6.000
x= 0.000       y=-2.000
x= 3.000       y=1.000
x= 5.000       y=13.000
```

[4]

```
%Problem Chapter 1 [4]
x=1;
while x<999
x=input('Input x value: ');
if x>=999, return
elseif x<0;
y=exp(x);
fprintf('x = %f, y=exp(x)= %f \n', x, y);
elseif x>0
y=1/(1+x^2);
fprintf('x = %f, y=1/(1+x^2)= %f \n', x, y);
end %if
end %while
```

Input x value: 9
x = 9.000000, y=1/(1+x^2)= 0.012195
Input x value: 0
Input x value: -9
x = -9.000000, y=exp(x)= 0.000123
Input x value: 999

[5]

```
% Problem Chapter 1 [5]
for N=[1000, 10000,100000];
v=rand(1,N);
count=0*(1:5);
```

169

```
for m=1:length(v)
bin=floor(v(m)/0.2)+1;
count(bin)=count(bin)+1;
end
f=count/N;
average=sum(f)/5;
variance=sum((f-average).^2)/5;
disp('  ')
fprintf('Number of random numbers N = %i\n', N)
disp('Fraction of N in each bin')
disp(f)
fprintf('Variance = %.3e\n', variance)
end %for
```

Number of random numbers N = 1000
Fraction of N in each bin
   0.19300   0.21800   0.18500   0.20400   0.20000
Variance = 1.228e-004

        Number of random numbers N = 10000
        Fraction of N in each bin
          0.19680   0.20180   0.20220   0.19620   0.20300
        Variance = 8.352e-006

        Number of random numbers N = 100000
        Fraction of N in each bin
          0.19839   0.20184   0.20124   0.19974   0.19879
        Variance = 1.809e-006

[6]

```
>> clear
>> a=hilb(5)
a =
    1.00000    0.50000    0.33333    0.25000    0.20000
    0.50000    0.33333    0.25000    0.20000    0.16667
    0.33333    0.25000    0.20000    0.16667    0.14286
    0.25000    0.20000    0.16667    0.14286    0.12500
    0.20000    0.16667    0.14286    0.12500    0.11111
>> a=a+1
>> save a
>> clear
>> load a
>> a
a =
    2.0000    1.5000    1.3333    1.2500    1.2000
```

1.5000	1.3333	1.2500	1.2000	1.1667
1.3333	1.2500	1.2000	1.1667	1.1429
1.2500	1.2000	1.1667	1.1429	1.1250
1.2000	1.1667	1.1429	1.1250	1.1111

[7]

u + v

2	3	4	5
3	4	5	6
4	5	6	7
5	6	7	8

c*v

5
8
8
16

c*w   Invalid operation

a+b

2	4	4
4	0	2
5	1	4

b*c Invalid operation

a*b

21	7	11
4	0	3
9	2	5

a.*b

1	0	3
4	-1	1
4	0	4

u.^2

| 1 | 4 | 9 | 16 |

[8]

>> a*b'   :Row vector times column vector becomes a scalar.
               Valid in linear algebra.

ans =   32

>>a'*b   :Column vector times row vector becomes a matrix.

Valid in linear algebra.

```
ans =
     4    5    6
     8   10   12
    12   15   18
>> a.*b   :Array arithmetic operation. Not in linear algebra.
ans =
     4   10   18
>> a.^b   :Array arithmetic operation. Not in linear algebra.
ans =
     1   32   729
>> a.*b'  :Array arithmetic operation. Not explained in text,
          but makes sense.
ans =
     4    8   12
     5   10   15
     6   12   18
```

[9]

%Problem Chapter 1 [9]	February Calendar
| ```
d=1:29;
d=[0,d];n=0;
disp(' February Calendar')
for k=1:ceil(d(length(d))/7);
for j=1:7
n=n+1;
if n>length(d); break; end%if
if d(n)<10,
fprintf('   %i', d(n));
else
fprintf(' %i', d(n));
end%if
end%for
fprintf(' \n');
end%for
disp('     ')
``` | ```
 0   1   2   3   4   5   6
 7   8   9  10  11  12  13
14  15  16  17  18  19  20
21  22  23  24  25  26  27
28  29
``` |

[10]

n	x	x^2	%Problem Chapter 1 [10]
1	0.0	0.000	n=(1:10)';
2	0.5	0.250	x=[0 0.5 1.1 1.6 2.2 2.7 3.4 3.9 4.4
3	1.1	1.210	5.0]';
4	1.6	2.560	xsq=x.^2;

```
5    2.2    4.840     fprintf(' n      x              x^2 \n')
6    2.7    7.290     for m=1:length(n)
7    3.4    11.560    if xsq(m)<10,
8    3.9    15.210    fprintf(' %i      %.1f           %.3f\n', n(m), x(m),
9    4.4    19.360    xsq(m))
10   5.0    25.000    else
                      if m<10
                      fprintf(' %i      %.1f           %.3f\n', n(m), x(m),
                      xsq(m))
                      end%if
                      if m>=10
                      fprintf('%i       %.1f           %.3f\n', n(m), x(m),
                      xsq(m))
                      end%if
                      end%if
                      end%for
```

[11]

```
%Chapter 1 [11]          Result:
A='Jopnes    ';          Z =
B='Smith     ';          Jopnes
C='Alexander';           Smith
D='Xing      ';          Alexander
E='Camden    ';          Xing
Z=[A;B;C;D;E]            Camden
```

Exercise problems for Chapter 2

[1]

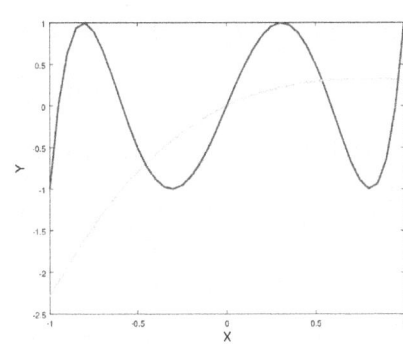

```
%Problem Chapter 2 [1]
x=-1:0.05:1;
y1=sin(x).*exp(-x);
y2=cos(5*acos(x));
plot(x,y1,'g', 'linewidth', 2);
hold on
plot(x,y2,'b', 'linewidth', 2);
xlabel('X', 'fontsize', 16)
ylabel('Y', 'fontsize', 16)
```

[2]

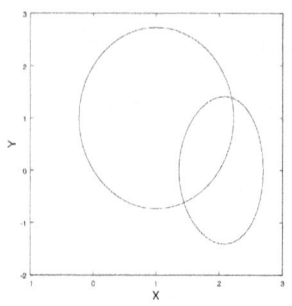

```
%Problem Chapter 2 [2]
clear all, clf
x1=-1:0.1:3;
y1=-2:0.1:3;
[x,y]=meshgrid(x1,y1);
f=2*(x-1).^2 + (y-1).^2 - 3;
contour(x,y,f, [0 0], 'k'); hold on
axis square
f=(x.^1.5-3).^2 + y.^2 -2;
contour(x,y,f, [0 0], 'k');
xlabel('X', 'fontsize', 16)
ylabel('Y', 'fontsize', 16)
```

[3]

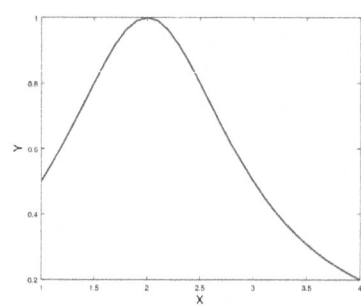

```
%Problem Chapter 2 [3]
clear all, clf
x=1:0.1:4;
y = 1./( 1 + (x-2).^2);
plot(x,y,'linewidth',2)
xlabel('X', 'fontsize', 16)
ylabel('Y', 'fontsize', 16)
```

[4]

Awful

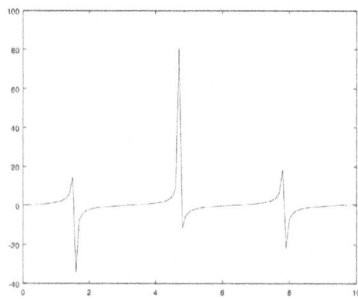

```
x=0:0.1:10;
plot(x,tan(x))
```

Awesome

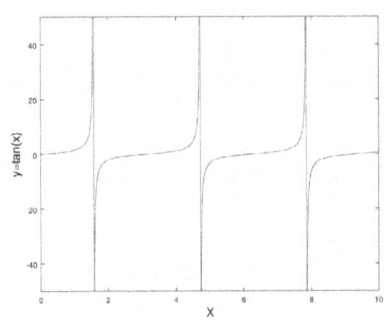

```
x=0:0.01:10;
plot(x,tan(x))
axis([0 10 -50 50])
xlabel('X', 'fontsize',
16)
ylabel('y=tan(x)',
'fontsize', 16)
```

[5]

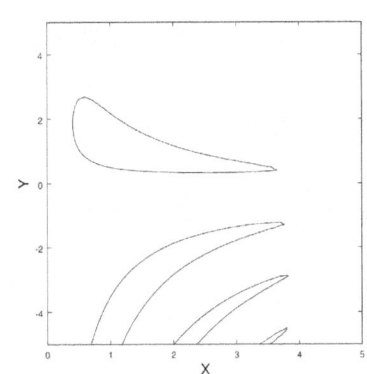

```
clear all, clf
x1=0:0.1:5;
y1=-5:0.1:5;
[x,y]=meshgrid(x1,y1);
f=sqrt(1+x.^2+exp(y))-4*sin(x.*y);
contour(x,y,f, [0 0], 'k');
axis square
xlabel('X', 'fontsize', 16)
ylabel('Y', 'fontsize', 16)
```

[6]

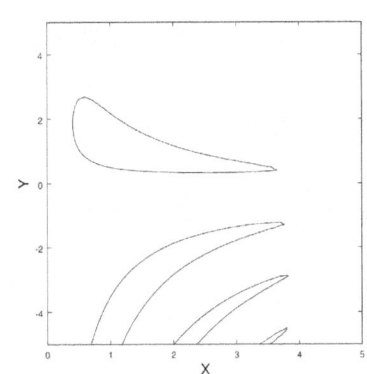

```
%Problem Chapter 2 [6]
clear, clf
x=-1:0.05:1;
y1=sin(x).*exp(-x);
y2=cos(5*acos(x));
plot(x,y1,'g', 'linewidth',2); hold
on
plot(x,y2,'b', 'linewidth',2);
xlabel('X', 'fontsize', 16)
ylabel('Y', 'fontsize', 16)
text(-0.6, 0.5, 'y=sin(x)exp(-
x)','fontsize', 14)
text(-0.9, -2,
'y=cos(5acos(x))','fontsize', 14)
```

[7]

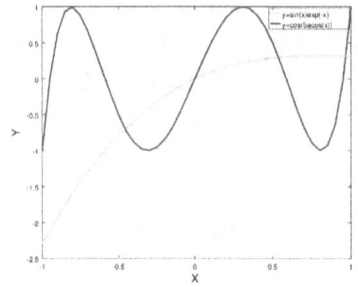

```
%Problem Chapter 2 [7]
clear, clf
x=-1:0.05:1;
y1=sin(x).*exp(-x);
y2=cos(5*acos(x));
plot(x,y1,'g', 'linewidth',2); hold on
plot(x,y2,'b', 'linewidth',2);
xlabel('X', 'fontsize', 16)
ylabel('Y', 'fontsize', 16)
legend( 'y=sin(x)exp(-x)',
'y=cos(5acos(x))' )
```

[8]

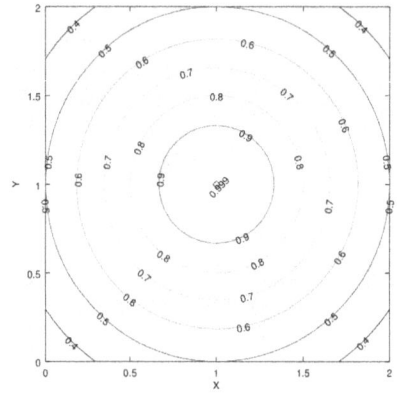

```
%Problem Chapter 2 [8]
x=0:0.05:2;
y=x;
[X,Y]=meshgrid(x,y);
f=1./( 1 + (X-1).^2 + (Y-1).^2);
level=[0.2:0.1:0.9,0.999];
h=contour(X,Y,f, level);
clabel(h)
xlabel('X')
ylabel('Y')
axis square
```

[9]

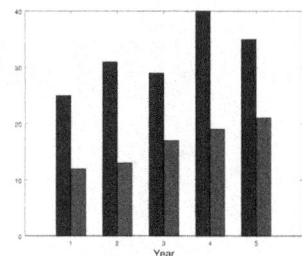

```
%Problem Chapter 2 [9]
data=[
1          25          12;
2          31          13;
3          29          17;
4          40          19;
5          35          21
]

bar(data(:, 2:3), 0.8, "hist")
xlabel('Year','fontsize',16)
```

Exercise problems for Chapter 3

[1]

ab=
    13   4
    -4   2
av=
    13
    14
inv(b)v=
  -0.71429
  5.57143

[2]

(a)   1, 1, 1   det=27
(b)   a=[-1 1 1 -1; 1 2 -3 3; -2 2 1 2; 1 5 -1 -3]
    s=[3 -8 -1 7]'
    a\s
    ans =
      2
      1
      3
      -1
    Det(a)=24
(c)   Coefficient matrix is singular, no solution.

[3]
```
%Problem Chapter 3 [3]
x=[0 1 2 3 4];
y=[0 1 0 0 0];
c=polyfit(x,y,4);
polyval(c,x);
disp('Coefficients of the polynomial')
disp(c)
disp('Verification of the polynomial')
disp(polyval(c,x))
```

Coefficients of the polynomial
  -1.6667e-001   1.5000e+000 -4.3333e+000   4.0000e+000   4.0852e-015
Verification of the polynomial
   4.0852e-015   1.0000e+000 -1.2438e-015   1.4207e-015   5.3252e-016

[4]

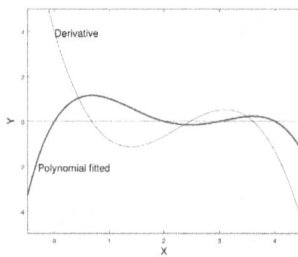

```
%Problem Chapter 3 [4]
clear,clf
x=[0 1 2 3 4]; y=[0 1 0 0 0];
c=polyfit(x,y,4);
polyval(c,x);
cd=polyder(c);xp=-0.5:0.1:4.5;
yp=polyval(c,xp);
ydp=polyval(cd,xp);
plot(xp,yp,'linewidth',2); hold on
text(-0.3,-2, 'Polynomial
fitted','fontsize', 16)
text(0, 4, 'Derivative','fontsize', 16)
plot(xp,0*xp); plot(xp,ydp);
axis([-0.5 4.5 -5 5])
xlabel('X','fontsize',16);
ylabel('Y','fontsize',16);
```

[5]

$y'= -1.05047$ $y''= 0.50122$ $y'''=6.26485$

[6]

```
%Problem Chapter 3 [6]
%Simpson's rule integration:
disp('S is user-definition of the integrant: example ')
S= 'f=log(1+x)./x; '
```

```
disp(' a, b, and n are user-input: following data are example
')
for n=[2, 4, 8, 16, 32, 64, 600];
a = 1; b=2; %n=30;
h=(b-a)/n; x=a+(0:n)*h;
eval(S);
disp('The result of integration ')
    I=sum(f)+3*sum(f(2:2:length(f)-1)) +
sum(f(3:2:length(f)-2));
    I=(h/3)*I;
    [n,I]
end%for
  h*( sum(f) -0.5*f(1) - 0.5*f(length(f)))
```

(a)

n	I
2	2.1376
4	2.1001
6	2.0978
8	2.0974
16	2.0973

(b)

2	1.2761
4	1.2874
8	1.2902
16	1.2910

(c)

2	0.61432
4	0.61428
8	0.61428
16	0.61428

[7]

User-definition of the equation to solve:
  f= sin(x).*exp(-0.4*x) ;

Itr.No	Itr. Sol, x	Residue
1	3.134901e+000	1.35e-001
2	3.141578e+000	6.68e-003
3	3.141593e+000	1.51e-005
4	3.141593e+000	-5.96e-009

End of solution

[8]

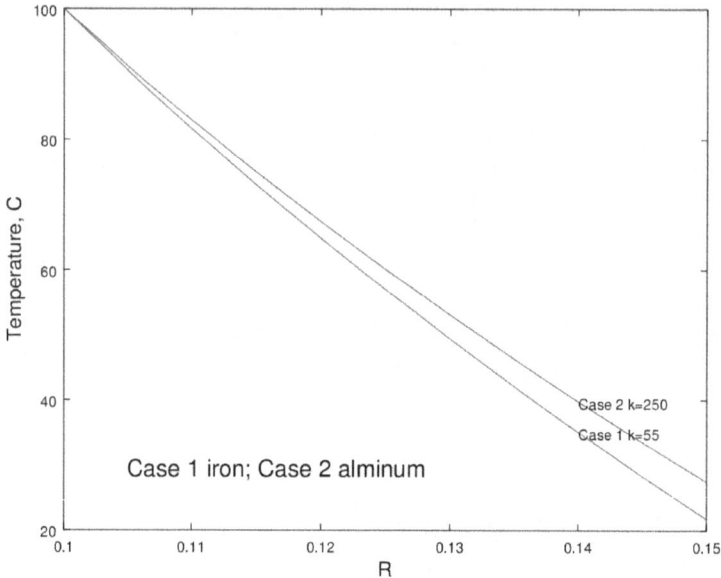

The surface temperature is higher for aluminum than iron. In general, as the thermal conductivity becomes smaller, the surface temperature becomes lower.

```
% Problem Chapter 3 [8]
clear, clf
Tair=20;
for kase=1:2
ht=200; % heat transfer coefficient
if kase==1, k=55;      %thermal conductivity iron
elseif kase==2, k=250;% thermal conductivity, alminum
endif
h=0.005;
r=0.1+h:h:0.15;
r(length(r)) ;
imax=length(r);
for i=1:imax-1
a(i)=-(r(i)-h/2);
c(i)=-(r(i)+h/2);
b(i)=-a(i)-c(i);
d(i)=0;
end%for
d(1)=-a(1)*100;
```

```
a(imax)=c(imax-1);
b(imax)=-a(imax)+r(imax)*ht/k;
d(imax)=r(imax)*ht/k*Tair;
%Tridiagonal solution begines
n=imax;
for i=2:n
   q=a(i)/b(i-1); b(i)=b(i)-q*c(i-1);
   d(i)=d(i)-q*d(i-1);
end %for
d(n)=d(n)/b(n);
for i=n-1:-1:1
   d(i)=(d(i)-c(i)*d(i+1))/b(i);
end %for;
% End of tridiagonal solution
T=d;
plot([0.1,r],[100,T]), hold on
text(r(imax-2),T(imax-2),['Case ',num2str(kase),   ...
      '    k=', num2str(k)])
end %for
xlabel('R','fontsize',14)
ylabel('Temperature, C', 'fontsize', 14)
text(0.105, 30, 'Case 1 iron, case 2 alminum','fontsize',16)
```

[9]
Properties of steel
Density of steel: ro=7800Kg/m^3
Specific heat: cp=0.49kJ/kgK
Thermal conductivity: k=40W/mK
Heat transfer coefficient in boiling water: 2000 W/m^{2}C

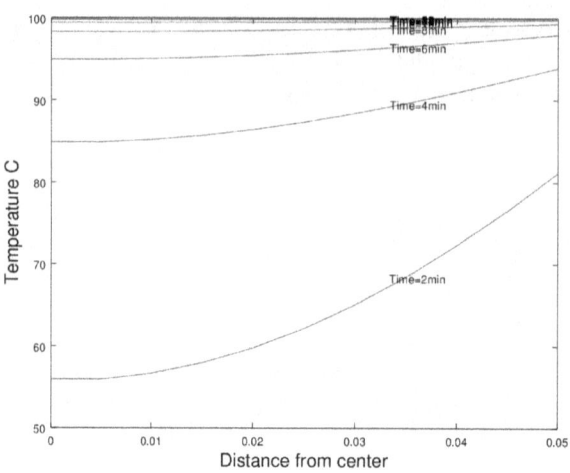

```
%Problems Chapter 3 [9]
clear, clf
k=40;          %W/mC
ro=7800;       % kg/m^3density
cp=490;        %J/kgC Specific heat of water
h=2000;        %W/m^2C    Moderaate speed
of air flow
R=0.05;        %diameter of cylinder, m
imax=10;       %maximum mesh index
Tair=100;      % degrees C
Tini=0;   %    Initial internal temperature
degrees C
dt=60;         %s time interval for calculatios
dr=R/imax;
rr=dr*(1:imax);
% Geometrical note I=0 is the center; i=1 is
next point not center
for i=1:imax-1
aa(i)=-k*(rr(i)-dr/2)/dr;
bb(i)=dr*ro*cp*rr(i)/dt+2*k*rr(i)/dr;
cc(i)=-k*(rr(i)+dr/2)/dr;
dcoef(i)=dr*ro*cp*rr(i)/dt;
endfor
bb(1)=bb(1)+aa(1);
aa(imax)=-k*(rr(imax)-dr/2)/dr;
bb(imax)=( 0.5*dr*ro*cp*rr(imax)/dt +
h*rr(imax)+k*(rr(imax)-dr/2)/dr);
cc(imax)=0; %dummy

%Continued to next column

time=0;
%Initial temperature condition
TB(1:imax)=Tini;
%for case=1:2
for timestep=1:50
time=time+dt;
a=aa;b=bb;c=cc;
for i=1:imax-1
d(i)=dcoef(i)*TB(i);
end
d(imax)=h*rr(imax)*Tair+0.5*dr*ro*cp*rr(imax
)/dt*TB(imax);
%Tridiagonal solution begins
n=imax;
for i=2:n
  r=a(i)/b(i-1); b(i)=b(i)-r*c(i-1);
  d(i)=d(i)-r*d(i-1);
endfor
d(n)=d(n)/b(n);
for i=n-1:-1:1
  d(i)=(d(i)-c(i)*d(i+1))/b(i);
endfor;
%end of tridiagonal solution
T=d;
if mod(timestep,2)==0
plot([0,rr,R],[T(1),T,Tair]); hold on
s=['Time=',num2str(time/60),'min']
rtx=R*2/3; Ttx=T(floor(rtx/dr)+1);
text(rtx,Ttx,s );
end %if
TB=T;
end% timestep
ylabel('Temperature C', 'fontsize',16)
xlabel('Distance from center','fontsize', 16)
```

[10]
Only next part changes from the program in Section 3.6. Computational results are shown in the two figures below. Temperature in sphere decreases become faster than in cylinder.

```
%Problem Chapter 3 [10]
% Geometrical note: i=0 is the center; i=1 is next point not center
for i=1:imax-1
aa(i)=-k*(rr(i)-dr/2)^2/dr;
bb(i)=dr*ro*cp*rr(i)^2/dt+k*(rr(i)-dr/2)^2/dr+k*(rr(i)+dr/2)^2/dr;
cc(i)=-k*(rr(i)+dr/2)^2/dr;
dcoef(i)=dr*ro*cp*rr(i)^2/dt;
endfor
bb(1)=bb(1)+aa(1);
aa(imax)=-k*(rr(imax)-dr/2)^2/dr;
bb(imax)=(                0.5*dr*ro*cp*rr(imax)^2/dt                +
h*rr(imax)^2+k*(rr(imax)-dr/2)^2/dr);
cc(imax)=0; %dummy
```

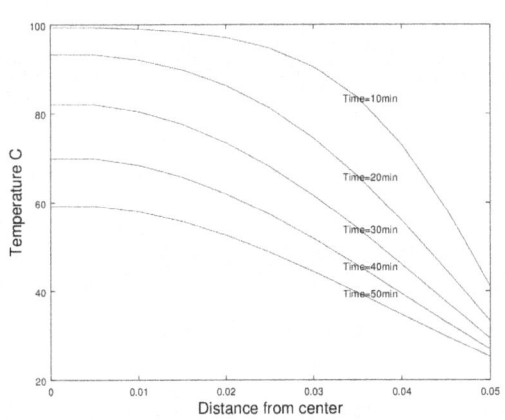

Case 1

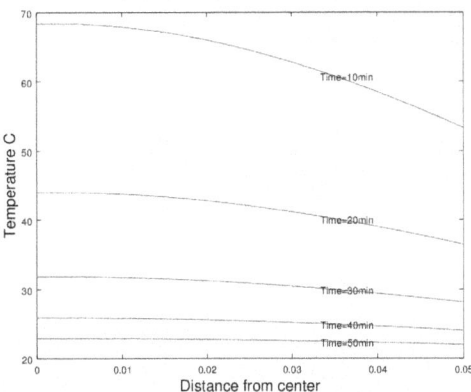

Case 2

183

References

S. Nakamura, Numerical Analysis and Graphic Visualization with Matlab, 2nd ed., Prentice-Hall, 2002

S. Nakamura, Foundation of Numerical Methods, Amazon.com, 2016

P.J.G. Long, Introduction to Octave, http://www-mdp.eng.cam.ac.uk/web/CD/engapps/octave/octavetut.pdf, 2005

Octave vs Matlab Image Processing, http://wizzcore.com/tag/octave-vs-matlab-image-processing

J. S. Hansen, GNU Octave Beginner's Guide, Packt Publishing, 2011

J. W. Eanon, GNU Octave Manual, Network Theory Limited, 2nd ed., 2005

Real Projective Plane: https://en.wikipedia.org/wiki/Real_projective_plane

Machinery's Handbook, 29th Edition Large Print & Toolbox Editions, http://www.engineersedge.com/heat_transfer/convective_heat_transfer_coefficients__13378.htm

Frequently asked questions (FAQ) for GNU Octave users, http://wiki.octave.org/FAQ#Compiler

GNU Octave/Mat lab Tutorial Series
Volume 1
OctaveMatlab Primer and Applications
EZ Guide to the Commands and Graphics
185 pages
Available at Amazon.com

Volume 2
Foundation of Numerical Analysis
Implementation with Octave/Matlab
164 pages
Available at Amazon.com

Volume 3
Numerical Solution of Ordinary Differential Equations
(Forthcoming, Amazon.com)

Volume 4
Numerical Solution of Partial Differential Equations
(Forthcoming, Amazon.com)